Chess Metaphors

Chess Metaphors

Artificial Intelligence and the Human Mind

Diego Rasskin-Gutman
Translated by Deborah Klosky

The MIT Press
Cambridge, Massachusetts
London, England

For information about special quantity discounts, please email special_sales@mit-press.mit.edu.

This book was set in Stone by Graphic Composition, Inc.
Printed and bound in the United States of America.

Library of Congress Cataloging-in-Publication Data

Metaforas de Ajedrez. English
Chess metaphors: artificial intelligence and the human mind / Diego Rasskin-Gutman; translated by Deborah Klosky.
　　p. cm.
Includes bibliographical references and index.
ISBN 978-0-262-18267-6 (hardcover: alk. paper)
1. Chess—Psychological aspects. I. Title.
GV1448.R37 2009
794.101'9—dc22

　　　　　　　　　　　　　　　　　　　　　　　　　　2008044226

Originally published as *Metáforas de ajedrez: la mente humana y la inteligencia artificial*, © Editorial La Casa del Ajedrez, Madrid, 2005

10　9　8　7　6　5　4　3　2

To Debbie, Gabriel, and Alexander

Contents

Foreword

I've been ensnared by chess during two periods of my life. The first time was when I entered the physics department as a university student. The second was last summer while I was pondering a metaphor from Richard Feynman comparing the observer of coffee-house chess games to a scientific researcher. In both cases, something similar happened: chess took over a large part of my conscious and unconscious thoughts.

In my youth, the boredom of classes (when the professor insisted on presenting things that could be read in books) pushed me into a nearby bar where speed games were played. At night, I would go to a club to help prepare games for the local tournament that was played on Sundays. After the games, I often dreamed about them: if I won, I dreamed that I lost; if I lost, I dreamed that I won. The brain has evolved throughout millions of years to anticipate uncertainty, and its joy lies between two limits — the offense when the uncertainty is too low (and the challenge is trivial) and the frustration when the uncertainty is too high (and the challenge is inaccessible). In that era, my mind was nourished more by coffee-house chess games than the canned science of university courses. But is it possible to speak of uncertainty in the case of chess? There are ten raised to the power of 120 different games. So, yes, it is possible to speak of uncertainty. Each player contains a ration of uncertainty for his adversary. The number of subatomic particles in the universe is on the order of ten raised to the power of eighty, and the number of different free sonnets that can be written in Spanish (fourteen verses chosen from among 85,000 words) is ten raised to the power of 415. In chess, the brain holds an illusion of creating, the same as when a poet writes a sonnet or when a scientist proposes a law of nature, taking the improbable chance that nature will accept it. In that period of my life, to create was to create chess games.

This last idea has to do with Feynman's metaphor, a brilliant idea that just a few months ago pushed me for the second time into the arms of chess. An onlooker of coffee-house games, who does not know the rules, can discover them if he observes a sufficient number of encounters. The metaphor, like all good metaphors, is rich in possibilities for thinking about the scientific method and philosophy of science—richer, even, than Feynman himself intended.

On both occasions, it was not easy for me to regain enough distance from chess to be able to concentrate on other parts of my daily tasks. But my friend Diego Rasskin-Gutman suddenly sent me the galley proofs of his last book. I opened the package. Danger. I recognized the sensation. I felt again that free-fall into the depths of chess. How is it possible that this many things about a wonderful game escaped me? How is it possible not to know about its historical origin as an oracle—that is, as an instrument for treating uncertainty? I had intuited many relations between chess and art, between chess and aesthetics (which is not the same), between chess and science, and between chess and intelligibility (which is also not the same), but it is one thing to intuit and another to know. I was lost. I looked for somebody with whom to play. I looked for memorable games from the past from my old myths—the brilliant Tahl, the astonishing Fischer, the legendary Capablanca, the amazing Najdorf, the solid Petrosian. I looked for someone with whom to talk about chess. But I was also struck by new preoccupations. Chess serves up metaphors, but it also serves up paradoxes, forms, methods, feelings, ideas, techniques, speculations, histories, indications, stimuli, intuitions, knowledge. This book by Diego Rasskin-Gutman is filled with all that. It gives the impression that the ideas in this work have been turning over and over in the mind of a scientist who is interested in everything that has to do with the complexities of the world while he thinks about science, plays chess, or contemplates a work of art. I will be lost in its ideas for a long time. And now that makes three—one captivity during my university days, another at the hands of Richard Feynman, and now yet another at the hands of Diego Rasskin-Gutman.

Jorge Wagensberg
Barcelona, May 2005

Preface

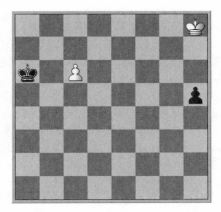

Gabriel looks at Alex. Alex smiles, provoking in his older brother a spontaneous laugh. Four-year-old Gabriel says, "Papi, Alex is little, and he doesn't understand what you say." After that, we read a book. He looks at me with admiration: "Papi, how can you do that?" "Do what?," I say with surprise. "That," he says. "How can you know the end of the story?" I look tenderly at him, as you do when contemplating the candid curiosity of a child, and gaze at Alex too, who is still smiling, caught up in the world of a just-about-one-year-old. The three of us laugh, and we wrestle on the sofa.

The same type of fascination that floods the mind of a child when he observes an adult performing trivial and everyday tasks with "such skillfulness" is at the origin of this book. I am talking about the sense of awe that I feel when following the games of the great chess players or the overwhelming emotion that I felt when suddenly I understood Richard Reti's composition (shown in the diagram above). I remember how on that occasion, I hardly had time to show my own father how white can avoid losing thanks to the geometry of the board (moving the king to the g7 square draws the game). But at many other times, I have been unable to comprehend all the subtleties that come in many of the moves, ideas, or strategic plans of the great players. In sum, I felt like my four-year-old son—in awe of a complexity (or simplicity) that escaped me.

My ability to know the ending of a child's story is something that few adults would consider surprising. In the same vein, the ability of a chess grand master (GM) to evaluate the position of the pieces on the board and predict where the pieces will be in the final position is a trivial matter for any expert player but not for those who don't know how to play chess or for players who do not have abilities beyond a certain level. The patzers who surround the chess tables of the local tournaments feel surprise after each move from the master—like children trying to understand how an adult knows that night follows day or that rain falls from the clouds. For most people, experts in any activity or human enterprise have abilities that are difficult to understand and easy to admire. But the mystery that envelops the enormous capacity of some people to carry out certain kinds of actions disappears with the acquisition of knowledge. We are all experts in something. Thanks to a long learning process, we can undertake incredible enterprises—from tying our shoelaces to solving differential equations. This capacity is the result of the never-ending curiosity of the human spirit, a curiosity that lies at the heart of this book.

How do we create this curiosity? Certain living beings, among them *Homo sapiens*, have a peculiar structure that we call the *brain*. Its presence allows them to explore the world through their five senses—touch, hearing, sight, smell, and taste. This happy union of perceptual and biological signal-processing elements is formed by an agglomeration of nervous tissue that, in turn, is formed by cells called *neurons* that specialize in the transmission of electrochemical information in the anterior end of the vertebrate animal (what we usually know as the head). The brain processes the stimuli that come from the outside world through our sense organs, elaborating pertinent responses that translate into action. To do that, the human brain employs remarkable resources. The memories stored in some cells of the nervous tissue are used as a comparative framework to the input stimulus, and the adequate integration of both inputs and framework is carried out in the form of thoughts that form meaningful responses.

For example, when we see a face, the visual information coming from our eyes stimulates certain areas of our brain, especially the occipital region (near the back of the neck) and the fusiform gyrus (in the temporal lobe), provoking a cascade of neural stimuli that elaborate amazingly varied kinds of thoughts—such as curiosity ("Where have I seen this face before?"), admiration ("What a beautiful face!"), a mundane greeting ("Hi, how are you?"), or caution ("I better go to the other side of the street"). But to remember faces, words, grammar rules, colors, or sounds, it was necessary to learn at some point in time. When we recognize, for example, the beauty of a face, we experience certain feelings and sensations all over our body that are difficult to describe with words: they are emotions. Moreover, when we recognize a face, we also recognize ourselves in the past and in the present in a brain exercise we call *consciousness*. All these brain activities—memory, thought, emotion, consciousness—are some of the many activities that are part of what we call *cognitive processes*. Western philosophy and science have always talked about them in a vague and loose manner as a unique concept—the *mind*.

In chapter 1 of this book, I introduce relevant aspects of the structure and function of the human brain to prepare a solid foundation for a discussion of the organic patterns and processes that underlie cognitive processes. I focus on the cognitive task of problem solving: which mechanisms operate inside our body when we encounter a problem that needs to be solved? These key issues of the biology of human behavior are explored, as are areas of study that have emerged around the science of artificial intelligence (AI). Certain intimate relations are highlighted when one looks at the cognitive properties of the brain within the context of theoretical modeling. Thus, AI is used here as a comparative model—a metaphor of the structural and functional organization of the brain. Along with AI, another metaphor—the behavior of a chess player—brings a narrower, more precise focus to the descriptions and ideas analyzed in this book. The moves of the thirty-two chess pieces over the sixty-four squares of the chessboard will guide this journey of discovery, opening up fundamental questions and several surprises.

The choice of chess as a common thread in this journey is not random. Chess offers an ideal lab for testing and exploring cognitive processes. Furthermore, the founders of AI as a scientific discipline (people like Edward Feigenbaum, Marvin Minsky, Allen Newell, Claude Shannon, Herbert Simon, Alan Turing, John von Neumann, and Norbert Wiener, among others) realized that chess offered a challenge of great theoretical interest to try to simulate thought processes through computer programming. So the core idea of this book can be encapsulated as follows: I use chess to delve into an analysis of the cognitive properties of the brain. As it explores concept after concept, move after move, the book delves into the varied mental mechanisms and the respective cognitive processes underlying the action of playing a board game such as chess. First, chapters 1 and 2 explore the basic organic elements behind the construction of a mind, and chapter 3 analyzes the formal contents of AI. Chapter 4 takes the game of chess into a bigger framework, analyzing collateral influences that spread along the improbable frontiers between games, art, and science. In chapter 5, I evaluate the results of AI's initial challenge—its search (with major programming efforts) for a computer that could beat chess grand masters and potentially become the next chess world champion. The book concludes with a critical analysis of such a challenge, examining why the real reach of this AI enterprise has been overstated when compared with the creative nature of the human brain.

The real world is not what it seems. This phrase, a common, everyday observation, encloses a deeper truth that suggests that accessing the world that surrounds us in a truly trustworthy way is impossible. The idea is part of the Western philosophical and scientific tradition, from Plato's cave to Heisenberg's uncertainty principle. Whether we are gazing at the long, flickering shadows of the allegorical cave or using tools that disturb the surrounding reality to analyze the minute particles that make up matter, we are condemned to hypothesize about the nature of the world. Worse still, according to Karl Popper, we are condemned to an inability to verify hypotheses. The problem of knowledge is how to formalize the possibility of recognizing an entity and then how to transmit its meaning in the best possible way. But knowledge just for oneself has little relevance. A few clues about the outer world are enough for survival, reproduction, and a peaceful and healthy existence. That is how most animals pass through the world, but almost all mammals and birds have elaborate behavioral organizations that lead them to form different types of associations. The human species is included in this group of social animals. As a part of our phylogenetic inheritance (*phylogeny* is the study of kinship relations among species with the implicit idea that these relations are due to evolution), the most complex social dynamics of evolutionary history have been generated—civilizations. Many invertebrate animals (like termites, ants, and bees) have also elaborated complex social relations, but they are far less sophisticated than the social hierarchies of birds and mammals.

Humans are, above all, cultural animals that have specialized in the acquisition and transmission of knowledge as if that were one more element of their biology. During the last thirty thousand years (at least), the human species has incessantly questioned its own nature and its position within the universe—an overwhelmingly empty enterprise because of the paucity of valid answers and always led by a search for religious meaning. Little by little, this search has been stripped of its divine sense as attitudes evolved through the influence of the scientific community and by the transformation of societies into modern nation-states. Thus, a huge role has been played by scientists.

Copernicus and Galileo showed that we are not the center of the universe, Darwin recognized the animal within our being, Freud placed consciousness at the center of the scientific quest, and Einstein equated matter with energy and showed nature's dependence on point of view. They helped generate a radical change in today's societies by promoting a spectacular turn away from the religious and toward the secular. But also important was a change of attitude that began with Enlightenment thinking in France and the United States of America at the end of the eighteenth century—a move toward the knowledge of reality and the conviction that the human being is but one participant in the world's natural landscape.

Chess is a cultural activity that originated somewhere in Asia in the second half of the first millennium CE. The next period of great development for chess was in the eighteenth and nineteenth centuries. Its growth paralleled the philosophical and scientific developments of the Enlightenment, especially in Central and Western Europe. It was perhaps the definitive thrust given by the cultural and artistic milieu of the nineteenth-century Romantic era that furnished chess with its present aura of intellectual and creative quality. Several cultural metaphors are embedded in chess—struggle as an echo of our animal nature (as the great Emmanuel Lasker put it), honesty, deceitfulness, bravery, fear, aggression, beauty, and creativity. Playing chess can resemble our attitudes about our daily lives and can sometimes take us beyond our personalities so that we can have on the board those dreams that the harsh reality of our lives forbids us. Chess is an activity that allows players to deploy almost all of their available cognitive resources. For this reason, chess is an ideal laboratory in which to start a journey into the diverse activities that are carried out by the mind and its physical correlates, including the ways that the brain functions and the body moves.

An analogy with biology can help to illustrate this last point. Research in biology is done with organisms that are considered models because they are easy to study for a variety of reasons. The data that we collect from these living models are generalized to other living beings. One of the most important of these models is the fruit fly (which has a more serious name—*Drosophila melanogaster*). Other important models are the bacteria *Escherichia coli*, the worm *Caenorhabditis elegans*, the frog *Xenopus laevis*, the chicken *Gallus domesticus,* and the mouse *Mus musculus*. The fruit fly has been cut in small pieces so that scientists can study the deepest corners of its anatomy, its sequence of embryonic development, the results of mutations in specific genes, and many, many other things. According to Neil Charness, chess is to cognitive science what *Drosophila* is to biology. Although Charness's optimism reflects more a wish than a reality, chess has been used for research in AI (which today is an undisputed branch of cognitive science) since its beginnings in around 1950. It is a model for emulating intelligent behavior through the use of algorithms and heuristics programmed in computer systems.

One of the first goals of the pioneers of AI was to create a program that could play chess. The idea, somewhat naïve, or at least, simplistic, was that if chess is considered an entirely mental activity, the emulation of the game by means of an algorithm would implicitly mean the simulation of thought production. For Claude Shannon, the father of information theory, the development of a chess program was not in itself useful but did offer an enormous gain in understanding of the heuristic needed to solve problems. This gain would be translated into a surge in automatic systems for handling and manipulating data and for practical situations in the world. The evolution of artificial intelligence as a research field shows that this move has been made from research on systems like chess to expert systems that have been used on a multitude of occasions to improve and automate tasks (from medical diagnosis to weather forecasting, two clearly useful examples).

From its beginnings in the mid-1950s, artificial intelligence as a scientific discipline has provoked debates about the nature of its name. Indeed, the idea that a kind of "artificial" intelligence exists seems somewhat absurd if one considers that the creation of this science is a cultural achievement of human beings: there is nothing artificial about it. But more important still is the definition itself of *intelligence* as a characteristic of the human mind that can be handled in descriptive and operative terms. The problems are multiple—from the possible existence of several types of intelligence (so that it might be possible to speak of multiple artificial intelligences depending on the type of intelligence that is being modeled or used in the simulator) to the conceptual impossibility of generating a behavior that can come close to intelligent human behavior (the problem of the semantic content of human knowledge versus the merely syntactic content of a computer program, made evident by John Searle in his Chinese room metaphor).

Why use models and simulations? The answer is simple: models are the foundation of knowledge. Every scientific theory is based on models of reality that approximate it to a greater or lesser degree. The first model of reality is language itself, where the symbolic capacity of humans to represent the world rests. It is not easy to break the barrier of language so that the knowledge generated about a given problem or parcel of the world escapes the restrictions imposed by words, as Michel Foucault masterly showed in *Les mots et les choses (The Order of Things)*. In ancient Greece, philosophers began to use models of nature that transcended words. These models were based on formal structures of diverse characteristics that constitute metalanguages of the representation of reality that have been preserved until today, although they have been so reworked that it is difficult to recognize them. The most important of these models are Pythagorean numbers, Aristotelian logic of predicates, and deductive Euclidian algorithms. Toward the end of the Renaissance, the numerical-mathematical models made their appearance, and thanks to their prediction powers, they fortified the development of science. Galileo, Descartes, Newton, Leibniz, and later, Ludwig Boltzmann, Einstein, and Turing (to name a few) used mathematical structures to model their parcels of reality. With Kurt Gödel, we find in the twentieth century the idea that formal systems are incomplete, a concept that is perhaps important to chess theory. If undecidable statements exist in chess, then it is impossible to solve them completely with a computer chess program.

Artificial intelligence falls within this scientific tradition, taking the field of modeling to a concrete parcel of the world—the brain. As such, it is a science that promises rewards within the history of human knowledge. Indeed, if we could manage to model the ways that we represent the world, put into practice our own capacity for introspection, and generate language, thoughts, emotions, and dreams—in short, that lengthy etcetera that we sometimes call *human consciousness*—it is likely that we would be able to finally approach answers to one of the deepest questions that the human species has always considered: who are we? In this sense, the main thesis that I propose in this book makes of chess a mental activity that can be modeled as a departure point for better understanding how the human mind works. The development of chess programs constitutes a field that has combined many of these perspectives—from the first attempts to emulate the mental processes of a game to the massive parallel searches within the tree of possibilities that are generated during play. The cognitive sciences (disciplines such as psychology, neurobiology, philosophy of the mind, and artificial intelligence) have influenced the systematic study of processes such as attention, knowledge, reasoning, logic, intelligence, information, memory, and perception, using the expert chess mind as a study metaphor.

On the surface, chess is a game that has a winner and a loser. However, a deeper look reveals that perhaps chess is not just a game but a line of communication between two brains. This hypothesis shapes the contents of the entire book: chess is a communication device. As with any other act of communication, it is necessary to have someone who sends the message, a transmission medium, and someone who receives the message. Players are both the communicators and receivers; the board and the chess pieces are the transmission medium. In an exchange of messages, ideas, attitudes, and personal positions about the uncertainty of our world, however, where is the win, and where is the loss? Another hypothesis complements the first one: human beings are always involved in situations where they must act as sources or receivers of information. Now the landscape begins to make sense. Perhaps chess is a game. But the life of a human being is also a game. All social activities in which a human being must become an information source and receiver have a series of factors in common. There is always something to communicate (moods, annoyances, happiness, feelings, ideas). There is always something that we need to understand (a noise, a color, a sign, a danger, an emotion). There is always some medium that is familiar to us (a grammar, an artistic language, a chess board).

The style of this book is personal. My scientific and academic background leads me to search for answers to every question and furthermore to look for the scientific answer. As a human enterprise, however, chess departs from science to reach into areas such as individual desires and social convention. I have tried to use language that is not too scientific to clarify some ideas. I also have tried to avoid bibliographic references. By noting just the names of certain authors, I hope to spare the reader never-ending parentheses that often are only obstacles to following ideas. Instead, a partially annotated bibliography lists the most important references that I used for each chapter. The first two chapters are the most biological ones and therefore the hardest for the nonspecialized reader. Chapter 1 introduces the biological bases of thinking, exploring the structure and function of the brain, while chapter 2 deals with contemporary ideas about the nature of the mind and cognitive processes, and chapter 3 presents AI as a brain and mental processes modeling tool. Chapter 4 is an introduction to chess as a human activity that justifies its choice as a model for the elaboration of cognitive theories, and chapter 5 describes the functioning of a chess computer program, its most relevant components, and the ways that the Internet is changing how people in the chess community relate to each other. After the epilogue, which includes some final reflections about the nature of chess, three appendixes present some important chess concepts, information about chess programs, and popular Internet sites. For the reader who is not familiar with chess, I recommend reading the appendixes first or consulting them whenever some concept related to the game is not sufficiently clear. Some sections of the book are necessarily technical, and readers should feel free to flip through parts that seem *too* technical, with the idea that reading should be a pleasure rather than a guilty obligation.

As a final word, I offer an answer to the following, quite natural, question: why have I written this book? Basically, I have put together in a single volume those ideas, elements, facts, visions, and surprises that I would have loved to find in a bookstore more than twenty years ago. I've tried to put chess within a framework that is larger than the game itself and that would help me to understand its reach as a cultural proposal, an elaboration of the mind, and a vehicle for understanding how the brain works. I have tried to relate the world that surrounds the effervescent task of playing a game of chess with human biology. I hope that the narrative and the metaphors that I have used transmit the fascination that I feel about chess to the readers who now have this book in their hands. If I have helped to reshape chess in new dimensions within the natural world and within the promising scenarios of AI, I have fulfilled my objective. The effectiveness of a gambit is always unknown. Only time tells players if they are right.

Diego Rasskin-Gutman
California, Spring 2005

Three years have passed since the first Spanish edition of *Chess Metaphors: Artificial Intelligence and the Human Mind* saw the light of day. The chess world, the AI world, and the brain/mind world have all evolved in new directions. Other books have joined this one, including *The Immortal Game* by journalist David Shenk, a stupendous account of chess, including a long section on cognition. Garry Kasparov has exited the chess scene to enter politics and has written *How Life Imitates Chess*. Viswanathan Anand is now the reigning chess champion. My sixteen-year-old nephew, Iván, drew him in a simul back in 2007 playing black with a king's Indian defense. I am proud of him. But what really caught my attention was an old book that I stumbled on two years ago—*Homo Ludens*, written in 1938 by the Dutch academic Johan Huizinga. This amazing book is an account of how games and playing have been culture generators throughout history, not just in obvious ways but rather as fundamental pillars that sustain the bases of societies. Although Huizinga did not include chess in his analysis, I was astonished at the similarities between the general mood of my book and Huizinga's thesis. If anything, it reassured me in my view that chess, as the noblest of games and as one nurtured by a plethora of cognitive processes, is to the human condition as brains are to the emergence of minds.

I am grateful to The MIT Press for providing the venue and the means to translate my book and especially to Robert Prior, who believed in this project from the beginning and helped me through the initial review process, and Ada Brunstein for her support and understanding.

Several things are different in this English edition. I have not included Adriaan de Groot's famous protocols, published in *Thought and Choice in*

Chess, because they can be easily accessed by the English-speaking community. Chapters 3 and 5 have been expanded with new sections on bioinspired strategies in AI and bioinspired approaches to chess computing. This last change was suggested by one of the reviewers, to all of whom I am also grateful.

The translation has been carefully crafted by my wife, Deborah Klosky. She has done a superb job of deciphering some obscure passages that not even I was sure about. My sons, Gabriel and Alex, are now seven and four, and Gaby knows the endings of many stories and most of the details of the seven volumes of Harry Potter, which he has read at least a couple of times. Now it is Alex's turn to wonder in awe about the same questions that Gaby had three years ago. We still end up laughing, and we wrestle on a (different) sofa.

Valencia, May 2008

Chess Metaphors

1 The Human Brain: Metaphor Maker

Over the lustrous green canopy of treetops that shade the heart of the Vienna Woods, where the Austrian Alps gently expire and meet a city full of history, nobility, and misery, black crows circle. They seem to give notice that not all is well—that these woods have seen terror and madness. In the late nineteenth and early twentieth centuries, everyone from every corner of the soon-to-be-dissolved Austrian-Hungarian empire could be found gathered together in Vienna—the Romantic writer who lived in those eternal cafés, his attention on the words he strung together and far away from the holes in his dirty, ill-fitting jacket; the musician who came from a village on the Italian border, Bratislava, or perhaps some forgotten town in Hungary to follow the same road as Mozart, Beethoven, Schubert, and Strauss; cultural and political figures who would change the course of history chose favorite cafés in which to nurture their dreams (musicians like Mahler and Schoenberg, scientists like Freud and Schrödinger, and politicians like Trotsky, Herzl, or the abominable dictator of Nazi Germany); and great chess players who like Wilhelm Steinitz and Carl Schlechter spent hours playing in the cafés, each one its own timeless universe, filling the chairs along with writers like Stefan Zweig, who captured the psychology of the chess player in his celebrated "Schachnovelle" ("Chess Story").

But in 1999, just before the arrival of the third millennium CE, the noise of the train that I was riding overpowered the caws of the black crows waiting in the distance to seize the souls of the dead, and my thoughts continued their own course—sometimes flitting from subject to subject, sometimes following a continuous stream. We arrived at Altenberg, my daily destination for two years. The Danube River runs parallel to the train tracks, and I saw a white swan swimming solemnly, reminding me of Konrad Lorenz and his noble intellectual misery. I started to walk toward the research institute, the Konrad Lorenz Institute for Evolution and Cognition Research, and as all the familiar sensations that I had experienced in recent days washed over me, an idea about how to solve a problem that I had been thinking about all week suddenly came to me as if out of nowhere. This chain of events is no mystery: it moved from my perception of the centuries-old trees to historical associations with the country's Nazi past, the consciousness of being in a train with its rhythmic sounds and movements, a swan on the river that suggested an image and a feeling, my own rhythmic body movement as I walked, and the sudden appearance of an idea completely unrelated to my other thoughts. No, it's no mystery: I am a mammal with a brain.

General Introduction to Brain Structure and Function

This chapter explores the biological bases for the development of cognitive processes in the human species. I use the expression *cognitive processes* to refer to those processes, either intentional or unintentional, that involve a certain type of stimulus-response mechanism. With such a broad definition, it is immediately clear that the majority of the poorly named higher animals show cognitive processes. That should not come as a surprise, given that all of them, including primates such as humans, possess a nervous system that has an accumulation of tissue in the anterior part of the body, which is organized into numerous specialized compartments. In other words, they all have a brain. Even creatures that lack a brain but are able to react to a stimulus from the environment and create a response accordingly possess cognitive properties. Bacteria in the microscopic world, for example, will respond to different concentrations of food in a water solution by moving to areas where the concentration is higher. In the plant world, the sunflower follows the direction of the sun during the course of the day. Humans have also created mechanisms whose functioning is based on pieces of knowledge, opening the door to the possibility that these machines also possess cognitive capacities. A thermostat turns on and off in response to the temperature around it, and a supermarket door opens when it perceives a customer's footstep.

In essence, this book is concerned with the following problem: what would we learn about our minds if we concluded that machines could carry out cognitive activities? And as a corollary to that idea, it also asks: can machines think? This extraordinary possibility has been the dream of many generations, and I return to it later to analyze it in greater depth. Before turning to the mind and cognitive processes, it is necessary to look at some of the basic characteristics of the brain—its structure and biological functions. With this biological foundation, it will be easier to relate cognitive processes (especially those that are commonly understood to be human) to the brain's functioning, which in turn is completely dependent on its form, its structure, and the connections between the elements that make up the brain—the neurons.

Form and Function: Brain and Mind

As with any animal organ, the human brain can be studied from both structural and functional perspectives. Each aspect has particular points of interest and needs to be looked at separately. As the great French anatomist Geoffroy de Saint-Hilaire pointed out in the nineteenth century, function must follow from the dictates of formal structure for any anatomical part. In other words, the functions that an organ can perform depend on the structural organization that it has (including its proportions, orientations, connections, and articulations) and the materials of which it is made. In engineering, this relationship is clear. Thanks to their particular structural conditions, a hanging bridge made of wood with tensors of rope can serve as a footbridge, while a bridge of reinforced concrete with steel tensors will allow cars to pass over.

In biology, the separation between form and function is a source of intense debate, especially in evaluations of the mechanisms that have played a role in evolutionary dynamics. Even so, it seems clear that the structure of the hand lets it carry out a variety of functions, from grabbing a rock to playing the piano. To assert that the hand has evolved to play the piano and thus has a structure developed in accordance with its function (in other words, that form follows function) is absurd (although appealing from a romantic point of view). The same thing is true of the brain. The extraordinary versatility of this organ is a direct consequence of how it is organized, and so its structural characteristics need to be closely examined. The eyes—the sight organs—have evolved structurally so that they can specialize in the reception of light stimuli. Again, function follows from form. Even so, form and function are integrated, meaning that in some cases it is hard to separate one from the other.

Structurally, the brain is a complex organ that is composed of billions of neurons and other auxiliary cells that together form an astronomical number of connections. Functionally, the brain is an organ that allows sensations from the environment to be evaluated, stored, and integrated and that provides appropriate responses to any given situation. To carry out these functions, the brain needs large quantities of energy. Although it makes up approximately 2 percent of the body's weight, it consumes up to 20 percent of the oxygen and glucose that are present in the blood, which is delivered through the blood vessels of the brain. The next chapter looks at the brain's functional capacities—that is, the brain as a process that commonly is called the *mind*—to create the worlds that represent, more or less reliably, the physical reality that surrounds it. The mind is a private witness of the course of our existence for each one of us. It is responsible for creating our emotional responses and for sensations like pleasure, happiness, fear, and hate. Or is it the brain that is responsible?

Some Structural Elements: Brain Cells

Since antiquity, the brain has been considered a continuous, undifferentiated mass of unknown matter that somehow (perhaps by hydraulic mechanisms, as Galen, René Descartes, and numerous other philosophers thought) used the nerves to send information to and receive information from the rest of the body. The nerves were considered to be hollow tubes through which a liquid transmitted the pressure that was sent by the brain. Thanks to Santiago Ramón y Cajal's anatomical and histological analyses, using silver staining techniques developed by Camilo Golgi, it became clear by the end of the nineteenth century that, like all other animal organs, the brain in reality is made up of a multitude of discrete elements called *cells* and that nerves are formed by a bundle of *axons*, which are just parts of those cells. These specialized brain cells, called *neurons*, are responsible for representing the world, storing memories, distributing information, and generating thoughts. Along with neurons, *glial* cells are a fundamental component of the brain. They nourish the neurons and isolate neuronal axons (much as if the axons were copper wires and the glial cells a plastic coating), facilitating the movement of ions through the wires. On average, there are about ten glial cells for each neuron, but since glial cells are about one-tenth the size of neurons, they take up roughly the same amount of space in the brain.

What is referred to colloquially as gray matter is actually those areas of the brain (principally those closest to the surface) that have a high density of neuron cell bodies and lack auxiliary glial cells, while the white matter of the brain looks that way because of the white color of *myelin*, the substance that sheaths the axons. Myelin is secreted by the *oligodendrocytes* glial cells (other types of glial cells are the *astrocytes* and *Schwann cells*). This sheath is an insulating material that is broken up at regular intervals, leaving gaps called *Ranvier nodes*. In these gaps, a large concentration of protein channels is situated in the plasmatic membrane of a neuron, which allows ions carrying an electrical charge to pass through the membrane (these ions or elements with an electrical charge are fundamentally sodium, potassium, and calcium).

The glial cells are important to the correct functioning of the brain, but here I focus on the neurons, since they are involved in generating cognitive processes. In addition to their classic functions of support, protection, control of pH in the environment, and nutrition (thanks to their intimate relationship with the blood vessels), many more functions for the glial cells have been discovered. This new, more active role for the glial cells includes influencing communication between neurons by regulating the ionic concentrations on each side of a neuron membrane. This makes the glial cells an indirect part of the processing of information, a function that has always been reserved exclusively for the neurons.

Butterflies of the Soul

Even so, the unquestionable protagonists of the brain are the neurons. These are specialized cells that assist in the reception, storage, integration, and distribution of the information that an organism encounters throughout its existence. Ramón y Cajal, who dedicated his scientific career to the study of the brain and was captivated by the complexity and delicacy of the external form of neurons, referred to them as "these butterflies of the soul." The human brain is made up of approximately 100 billion neurons (the exact number is unknown, but this as reasonable an estimate as any other). It can be thought of as a data center that, in an orderly fashion, receives, stores, integrates, and transmits information in electrochemical form. The cells' shape and size fall within a fairly wide range of variation. They interact among themselves by connections between their entrance and exit structures (their *dendrites* and *axons*, respectively) across connection elements called *synapses* (a synapse resembles a button and acts as a plug or outlet, to use the metaphor of an electrical circuit, bearing in mind that synapses are actually much more complex). Each neuron has connections to approximately 10,000 other neurons, so that the final result—the connectivity structure of the brain—contains an extremely high number of connections.

The basic morphological structure of a neuron consists of three distinct parts—the neuron body, the axon, and the dendrites (figure 1.1). The neuron body houses the cell nucleus as well as various *organelles* (such as many energy-generating *mitochondria*). The axon, a filament of varying length (from a few microns up to some 150 centimeters), provides a transmission path where the *action potentials* (the electricity that is generated by the difference in charge among chemical ions, especially sodium and potassium) travel and transmit a stimulus to other neurons. Finally, the *dendrites* are filamentous structures that are shorter than axons but are present in large

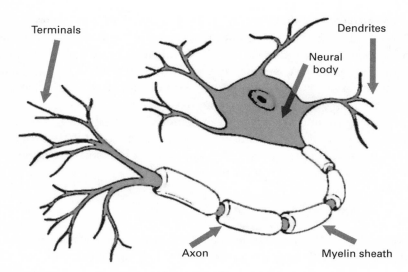

Figure 1.1
The main parts of a neuron.

numbers. Dendrites receive stimuli from the axons of other neurons. Other important elements in the neurons that should be mentioned are the *vesicles* (*microsomes*), which carry the chemical messengers (such as *neuropeptides*) that the dendrite needs to communicate with the axon; and the *membrane channels*, which are formed by diverse types of proteins and provide an entrance and exit for the chemical ions that assist in electrochemical current transmission.

There are also internal structures called *microtubules* that are found in the interior of the majority of animal cells. These filamentous structures (which are formed mostly by proteins like tubulin) confer a certain stability on the form of the neuron branches and are involved in the transport of *neurotransmitters* from their production site in the *soma* or cell body to the synaptic terminals. Roger Penrose has postulated that microtubules and information transmission among neurons are somehow related by an essentially quantum mechanism. This fascinating theory assigns a critical role to the microtubules, to the point of considering them necessary for the emergence of consciousness.

Neuron connections among the distinct areas of the brain are established during its embryonic development, partly in accordance with the spacing directives of the glial cells. In addition, specific types of neurons with distinct kinds of biochemistry, physiological conditions, or connection types

Figure 1.2
Some types of neurons—Purkinje cells, pyramidal cells, and star-shaped cells.

become differentiated from each other. Among the neuron types that are differentiated by their morphology are the pyramidal cells, Purkinje cells, motor cells, and star-shaped cells (figure 1.2).

Besides this morphological division, types of neurons also become differentiated by the kind of information that they are able to process. Thus, motor neurons, sensory neurons, and interneurons are formed during *embryogenesis. Motor neurons* stimulate movement in some part of the body, such as the neurons that innervate the muscles that move the fingers. The bodies of these neurons are lodged in the spinal cord, but their axons can be of enormous length (up to a meter and a half) to reach the indicated part of the body (for example, the big toe). *Sensory neurons* receive information from the sense organs (touch, hearing, sight, smell, and taste) and translate (the scientific term is *transduce*) the physical stimulus that arrives through each organ's specialized receptor cells into the electrochemical signal that neurons transmit to specialized regions of the brain. In the skin, for example, the receptor cells are the nerve endings themselves, while for vision, cells (*retinal cones* and *rods*) send the information from the luminous impression to the sensory neurons' terminals. The study of how vision is structured and functions is a productive topic in neuroscience and is providing a large quantity of information about the overall functioning of the brain and the nature of consciousness. Finally, the *interneurons* are neurons that play the role of intermediary between the sensory and motor neurons. During embryonic development, they establish routes where the sensory neurons transduce the information received from the environment and transmit it to the brain. There, the information is integrated and processed, and a new type of signal is produced that heads to a motor neuron, which provokes some kind of movement or action. But there are also reflex actions that establish a circuit (the *reflex arch*) that does not pass through the brain so that the action is much faster and unconscious.

Some Functional Elements: Electrochemical Current and Communication among Neurons

The brain is like a huge, highly complicated electronic circuitry where axons and dendrites serve as the wiring along which the electrochemical current travels. Current is generated by means of action potentials, and axons and dendrites communicate chemically by means of the release of chemical messengers called *neurotransmitters*. These two functional aspects create an electrochemical language of incredible reach.

An action potential is a process that releases a flow of electrical energy, produced by the depolarization of the membrane in a part of the neuron (figure 1.3). This depolarization is carried out by a series of protein channels in the membranes that allow certain chemical ions to pass between the interior and the exterior of the cell. Thanks to the delivery of neurotransmitters, the presynaptic neurons stimulate other neurons (the postsynaptic), producing a depolarization in the plasmatic membrane of the postsynaptic cell, which then evaluates all the individual contributions of the dendrites on this portion of the neuron to determine whether to fire (carry the impulse forward) or not. To a certain extent, this is a statistical problem, where the probability that the postsynaptic cell will fire is directly proportional to the sum of the excitations and inhibitions of all the afferent dendrites (dendrites coming from other neurons). In a resting state, the interior of the neuron holds a charge of some -70 millivolts with respect to its exterior, and when it allows sodium ions to pass through, the interior charge changes to 0 millivolts; that is, it depolarizes. A different phenomenon, that of *hyperpolarization*, happens when the difference in potential increases.

A close look at the structural and functional details of neurons shows that the wiring metaphor does not hold up. Axons receive and transmit action potentials in stages, given that the axon's conductance is small and the potential needs to be generated at regular intervals to be transmitted long distances. Moreover, the action potential is not transmitted directly from neuron to neuron (as happens with regular electric wiring) but instead requires the involvement of intermediary molecules that assist the transmission by leaving signals that change the possibilities of the action potential continuing in another neuron. (In certain cases when the synaptic gap is very narrow, the potential can jump from the pre- to the postsynaptic neuron.)

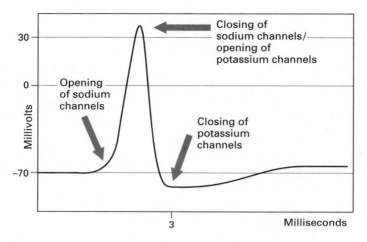

Figure 1.3
Action potential. The membrane of the neuron depolarizes, generating an electrical current with a characteristic transmission provoked by an ion flux.

This last aspect of neuron functioning is important because it constitutes the basis of cerebral functioning. Each axon possesses collateral branches where the action potential is distributed. These collateral branches connect (make synapses with) approximately 10,000 dendrites of thousands of neurons. For a neuron to be stimulated and transmit an electric current through its axon and its collateral branches, its firing *activation threshold* needs to be triggered. All the dendrites' contributions are added up to evaluate whether the neuron has passed that threshold. The sum is determined by the number of excitatory and inhibitory stimuli that are received, which in turn are determined by the type of neurotransmitter that is released. Neurons' electric functioning is also distinguished from the wire metaphor because in neurons the current moves by means of pulses. Once the neuron has crossed the activation threshold and thus fired its own action potential, it needs a period (the *refractory period*) of between 200 and 500 milliseconds to recover before it is ready again for the next firing.

One phenomenon that should be noted (and that could be behind the generation of high-level cognitive activities and even the manifestation of consciousness) is the phenomenon of *binding*, by which a group of neurons fire at the same time. This synchrony of neuron firing can be recorded using some of the methods discussed below (such as inserting microelectrodes in specific areas of the brain). Binding, or the rhythmic activity of neuron groups, could be what enables cognitive functions (such as perception and thought) to be carried out. To experience a scene consciously, its different properties need to be united in a coherent manner. This can be achieved through the neurons' action potential thanks to the phenomenon of binding. For example, when a person observes the ocean in movement, the brain must correlate different perception events (such as the color of the water, the smell of iodine, the movement of the waves, the reference to the horizon, and the sound of seagulls) and any other circumstantial phenomenon (such as the smell of sardines on the grill being prepared to end the day at the beach). All these events, sensations, perceptions, and feelings should be coded and represented in the brain as a unit, forming the conscious experience of a pleasant day at the beach.

The same thing happens within the conscious experience of a chess player. Many experiences—the color of the board, the texture of the pieces, the sound of the clock, the mental struggle to match up the positions on the board with memories of other games—should be linked with a physical presence in a specific part of a specific city and the tumultuous reality of the exterior world. All these elements together form a single, vital experience that is continually being processed, consciously or unconsciously, in the player's mind. And for that to happen, this phenomenon of firing synchrony among neurons from different parts of the cerebral cortex could be key.

Information is distributed in the brain in such a way that serial connections are scarce (and perhaps nonexistent within the brain, given the number of collateral branches of axons that also distribute information) compared to parallel connections. This means that electrical activity is spread throughout different brain areas. Neurons form particular networks that depend as much on development as on learning. These networks have specific, although not static, connection patterns and in many cases (as is shown below) can be identified with certain functional modules or parts of modules.

Depending on the type of neurotransmitter that is released in the synaptic button, the connection might contribute to the postsynaptic neuron's excitation or inhibition. This type of regulation of neuron communication (known generically as *neuromodulation*) is of great importance. Thanks to the differential action of distinct types of neurotransmitters, the brain can enter into different conscious states, such as dreaming or waking consciousness. Broadly speaking, we can distinguish between the *cholinergic* and *aminergic* systems, groups of neurotransmitters that respectively augment or inhibit neurons' excitation. Neuromodulation is important in bringing about different states of consciousness and also contributing to creating memories at a cellular level and therefore as a base for learning and memory.

Cellular Mechanisms for Memory Storage

Memory is a cognitive process that is necessary for learning. It is the basic component of intelligent behavior. The brain needs to possess a means for storing and recovering information that is registered by the sense organs or created by thought. This is achieved thanks to different memory systems, such as working memory and long-term memory (see the next chapter). However, some kind of molecular or cellular mechanism must allow that storage to take place. And this can take place only within the neurons.

In the 1940s, Donald Hebb proposed a mechanism for reinforcing the connections between two neurons (*neural memory*) based on the need for both the presynaptic neuron as well as the postsynaptic neuron to activate (*fire*) in a congruent or associative manner (when one fires, it increases the probability that the other will fire). This supposes a metabolic restructuring of both cells, strengthening their relation. Another mechanism that allows for learning at the synaptic level is the modulation of a third neuron that acts to reinforce the synapse between the pre- and postsynaptic neuron. In this case, the modulating neuron reinforces only the presynaptic neuron's activity.

One type of neuron connection is directly related to the generation of memories and learning. A special membrane channel called *N-methyl-D-aspartate* (*NMDA*) depends for its activation on the presence of the amino acid glutamate (which acts as a neurotransmitter), on the depolarization of the membrane, and on the stimulation of the neuron by a second path. The NMDA channels contribute to establishing what is called *long-term potentiation* in those synapses where they appear. They have been observed to be particularly numerous in the membranes of neurons in the *hippocampus*, a region of the brain that is related to memory storage. The receptor, situated on the postsynaptic membrane, receives an electrical signal thanks to the depolarization of the membrane, which happens independently because of the action of another type of postsynaptic receptor. This depolarization induces the NMDA receptor to release magnesium and allows it at the same time to bind with glutamate, which prompts calcium to enter the cell. The entrance of calcium seems to provoke a chain of reactions that culminates in the release of nitric oxide, a gas that acts backward as a messenger from the postsynaptic to the presynaptic neuron. There is a feedback mechanism between the two cells, mediated by the glutamate and nitric oxide, that strengthens the synaptic relation (or even stimulates a greater response). This constitutes a cellular memory mechanism, and thus the name of synaptic long-term potentiation.

Basic Organization of the Brain's Functional Regions

Higher vertebrates' nervous systems are divided into central and peripheral systems. The *central nervous system* (*CNS*) is composed of the brain and the spinal cord. The brain is lodged in the head and is protected by the braincase and by a set of flat bones connected to each other by strong, rigid sutures. The brain is also protected by three membranes—the *dura mater, pia mater,* and *arachnoid.* These membranes (also called *meninges*) protect the cerebral mass from infections and from knocks against the inner part of the cranium bones. The brain's average cellular mass weighs approximately 1.3 (for women) and 1.5 kilograms (for men). The spinal cord also has a set of bones that protect it (the *vertebrae*) and that are distributed along the cord's length like rings placed one on the other. The nerves that communicate between the brain and the rest of the body through the spinal cord pass through the intervertebral spaces, which are protected by cartilaginous disks. These nerves form the peripheral nervous system and, connected to the spinal cord, reach every part of the body. The nerves collect information from both outside and inside the body, which is then processed in the brain to elaborate an appropriate response. This response can be of a motor type (a movement, for example of a hand moving to swat a mosquito) or not (a thought). Furthermore, control of vital organs is carried out automatically, also thanks to the coordinated action of nerves and muscles.

The brain possesses various regions that are anatomically or functionally delimited (figure 1.4). The most basic parts are the hindbrain, the midbrain, and the forebrain (the latter provides the well-known image of the brain, the *cerebral hemispheres*). The *hindbrain* (also called the *lower brain*) consists of the *brain stem,* which connects the brain to the spinal cord; the *cerebellum,* a singular structure that is immediately posterior to the brainstem; the *medulla*; the *pons*; and the *reticular zone.* Functionally, the hindbrain structures control the body's vital functions, such as breathing, heartbeats, and digestion. They are also responsible for coordinating body movement, especially the cerebellum. The pons receives information from the visual areas, and the reticular formation controls the passage from sleep to wakefulness. The midbrain rests on the lower brain and is divided into the *tegmentum* and *tectum* (with two zones, the *inferior* and *superior colliculus*). The midbrain controls motor activities and sight and hearing (although in humans, both the visual and auditory areas are found principally in the forebrain's *neocortex*). The *forebrain* (or *upper brain*) is highly developed in humans. It is formed by hemispheres with *convolutions* (also known as *sulci* or *fissures*) and *grooves* (also known as *gyri*) with important internal structures. The extensive development of the forebrain (especially the neocortex) provides the structural conditions for the high-level cognitive functions to emerge. Other structures that are found in this area of the brain include the *thalamus* (the center of coordination for various sensory areas); the *hypothalamus* (which controls primary activities like feeding, flight, fight, and sex and regulates body temperature, sleep, and emotions); and some structures of the *limbic system* (the principal center of emotions control, along with the hypothalamus), such as the *hippocampus* (essential for fixing recently formed memories), the *pineal gland* (the seat of the soul, according to Descartes), and the *basal ganglia* (responsible for motor control).

All these structures are found in the interior of the forebrain. But the structures that have had the most pronounced development in the human brain are in the *neocortex*, which lies over the rest of the brain, covering the other interior structures. The neocortex, some two millimeters thick in humans, is divided into hemispheres (left and right). The *central longitudal fissure* lies between the hemispheres, which are connected mainly by a central structure called the *corpus callosum*, although there are other connection paths that do not pass through it. The most notable characteristic of the cerebral hemispheres in humans is their rugged appearance. Their folds and grooves provide a substantial increase in surface area and thus accommodate a greater number of neurons. Each hemisphere controls the information input and output corresponding to the opposite side of the body. Thus, the right hemisphere receives stimuli from the left hand, while the left hemisphere receives them from the right hand. Externally, the

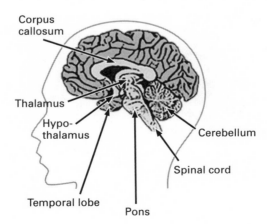

Figure 1.4
Sagittal section of the brain showing internal parts.

neocortex is divided into a series of regions bordered by fissures. The frontal and parietal regions are separated by the central fissure, while the lateral fissure separates those two regions from the temporal and occipital regions (figure 1.5).

The neocortex is diverse functionally and is where the high-level cognitive functions originate. For example, the frontal lobes (whose development has been more pronounced than the rest of the neocortex) carry out planning functions and possibly are the site of long-term memory storage, while the lateral lobes seem to be involved in decision making. In reality, the connections among the different areas, not only within the neocortex but also in the rest of the cerebral areas (especially the hypothalamus and basal ganglia), allow different cognitive elements to be integrated, producing the factors necessary to generate mental activity.

Over the past 150 years, researchers have been identifying areas of the brain. From this research, an image of the brain as a modular structure has emerged, with each module carrying out a specific activity. But the modularity of the brain has often been exaggerated, especially when considering the generation of cognitive processes. For example, pseudosciences like phrenology understand the brain as having specific parts for each cognitive process (even proposing modules for areas like friendship and morals) and also suggest that the skull's external, anatomical structure itself, with its indents and bumps, indicates the degree of development of these modules. In a less spectacular version of phrenology, Jerry Fodor and other psychologists and philosophers of the mind, using the Swiss army knife as a metaphor, insist that structural and functional modules for each type of action must exist.

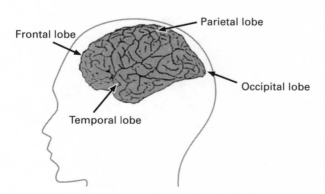

Figure 1.5
The brain in lateral view, showing the most important areas of the neocortex. Compare with figure 1.4.

They view the brain as an organ that is made up of multiple parts, each one of which is in charge of a series of specific tasks (figure 1.6). The idea of the modularity of the brain also has interesting consequences from an evolutionary point of view, given that evolution can simply add modules to those already there to generate a brain with new capacities.

Despite controversies about the modular brain hypothesis (at least in its most radical versions), the existence of specialized areas to receive particular stimuli is undeniable. This specialization is carried out during development and is largely provoked by the specific input of sensory information from both the body itself and from outside it. Some of the important areas of the neocortex are the visual areas (principally those called *V1* to *V5*), the motor areas, and some specialized areas such as those for language (Broca and Wernicke), color, face recognition, and memory (figure 1.7). Perception of an object is separated into attributes like movement, color, form, and orientation, given that each of these aspects is processed separately in different visual regions of the neocortex. Some areas of the visual cortex are organized in a way that represents the stimulus in a topologically equivalent manner. This means that if we could watch the activation of the neurons in those areas, they would form an image that was basically the same as the perceived object. However, other areas work in a much more diffuse way, and the equivalence is lost.

In the *motor cortex*, which is located in the most posterior part of the frontal lobe, the body is represented with an almost topological equivalence, and the same is true in the *somatosensory cortex*, which is just behind the motor cortex in the most anterior part of the parietal lobe. The most interesting part of these projection areas is that although the body's topological

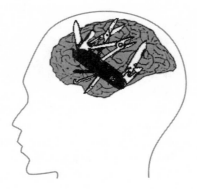

Figure 1.6
The brain as a modular complex, like a Swiss Army knife.

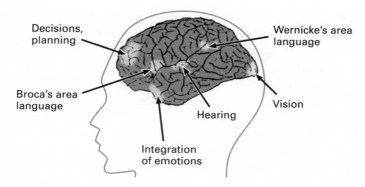

Figure 1.7
Some of the brain's functional areas. Even though the existence of these modules has been sufficiently demonstrated, cognitive functions need the joint action of several areas of the brain.

representation is almost equivalent (after the hand comes the forearm, which is followed by the upper arm, and so on), the relation between the real size of each body part and the size or proportion of the area as it is represented is completely different. Thus, the large area dedicated to the face and the hands stands out, reflecting the control of the many functions that these body parts can carry out. The relation between the motor and somatosensory cortices and the body is often represented by drawing a distorted human figure (called *motor homunculus* and *sensory homunculus*) on a section of each brain area (figure 1.8).

Besides the functional areas, which are concentrated in specific sites of the brain, the neocortex is stratified, with a cross-section showing six morphologically distinct layers with different connectivity patterns. These six layers, which extend across the whole surface of the neocortex, also have distinct specializations from a functional point of view. For example, layer II is composed mostly of pyramidal neurons with far-reaching axons, which make synapses with nonadjacent zones.

For an organism to survive, it needs diverse regulatory systems that are independent to a certain extent from the brain's integrated control. Thus, various systems contribute to keeping the body's interior in a dynamic equilibrium that is maintained always within a constant range for properties such as temperature (around 37 degrees Centigrade in adult humans) and pH (the ionic concentration that determines the acidity of the blood and other liquids inside the body and that varies from one organ to another; for example, the stomach pH is much more acid than the blood pH). This capac-

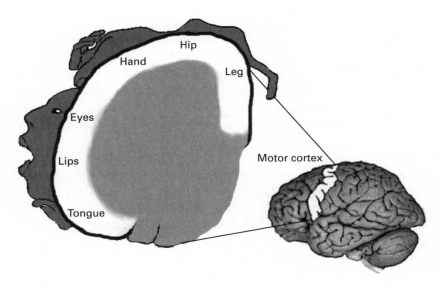

Figure 1.8
Spatial representation of body parts on the motor cortex. This topological relation between body parts and the brain has been traditionally drawn as a homunculus around the transversal section of the motor cortex. This kind of diagram shows that some anatomical parts are more represented (and hence controlled to a greater extent) than others in the motor cortex. Hands, which have built civilizations, are a significant example.

ity of the body to maintain a dynamic equilibrium is called *homeostasis* (the mid-nineteenth-century French physiologist Claude Bernard discovered the existence of this self-regulated *milieu interieur*, or internal environment). Another type of important regulation affects the automatic movement of muscles that control the body's vital organs. The brain's conscious control is not needed for the heart's *systolic* and *diastolic movements* (contraction and expansion) or for the diaphragm's movements that make breathing possible. These movements are controlled by the *autonomic nervous system*, which is divided into *sympathetic* and *parasympathetic systems*. The autonomic nervous system starts being formed during the first phases of embryonic development, and any defect in it has catastrophic consequences.

Another fundamental system of internal control is the *endocrine system*, a collection of organs that are located in different parts of the body and that secrete chemical substances (*hormones*). The pituitary gland or hypophysis, the thyroid gland, the adrenal glands, the ovaries, and the testicles are examples of hormone-producing organs that contribute to the regulation of vital functions. The brain and the hormonal system are closely related. A hormonal gland (the *pituitary*) is lodged within the brain. This gland is directly related to the hypothalamus, creating what is called the *hypothalamic-hypophyseal axis*. This axis controls all the vital processes that are carried out in the organism (such as breathing, heartbeat, blood regulation, and body temperature). The relation between the hypothalamus and the pituitary gland also regulates many basic behaviors (such as hunger, thirst, fear, reaction to cold, and sexual desire).

Although the brain is essentially not involved in the control of the regulation of the autonomic systems, there is a close relation between the autonomic systems and certain areas of the brain such as the limbic system, which is involved in the control of emotions. In this way, the basic physiological functions are directly connected to what we feel in a given moment. This point is fundamental for theories of consciousness and of formation of the mind because at a certain level, it is impossible to separate autonomic activities from conscious activities. And likewise, the vital functions can conclusively determine the kind of behavior that we carry out in a specific situation.

The brain's organization is separated into regions with structurally and functionally delimited areas, and it responds to criteria that are related to the laws of *bilateral symmetry*. This principle of organization is one of the fundamental characteristics of animal architecture. The phenomenon by which functional symmetry is broken in the brain hemispheres is called *lateralization*.

Vertebrates are organized according to principles of bilateralism. Fundamentally, *bilateralism* refers to the process of development by which the embryo grows in a symmetrical manner, which is a way to conserve resources during *embryogenesis*. These development processes are so fundamental that they are shared by a large group of animals (as disparate as worms and humans) called *Bilateria*. In an adult animal, bilateral symmetry can be easily recognized because the parts of the body on each side of an imaginary middle line are mirror images. The clearest example is the organization of the skeleton in vertebrates, which possess bilateral symmetry with respect to the middle axis that passes through the spinal cord: the hands, arms, legs, and ribs on one side of the body are the mirror image of those found on the other. Additionally, organs such as the kidneys and the cerebral neocortex (which has two hemispheres) fulfill the criteria of bilateralism. The heart, stomach, intestines, liver, and pancreas do not follow these rules. They are unmatched organs that lie on one side or the other of the body.

In spite of the symmetrical organization of vertebrate bodies, left and right differences can be seen even in the skeleton. The length of the legs, for example, is a well-known example. The same thing happens with the brain. Each cerebral hemisphere possesses a series of modules that are functionally distinct to such an extent that sometimes a person is spoken of as having a left- or right-dominant brain. The right hemisphere has been shown to specialize in spatial representation, among other cognitive activities, and the left to contribute to the understanding and expression of language. This difference could be behind a person's capacity to become a grand master in chess and the differences between men and women in the practice of the game. There are no conclusive studies in this respect, but activities such as mathematics, music, painting, architecture, and even chess, which require good spatial representation, are chosen by more men than women.

On the other hand, instead of an innate difference based on the asymmetric structure of the cerebral neocortex, these differences could reflect differences in education that foster the appreciation of certain activities above others. Thus, in chess, the appearance of the three Polgar sisters (Susan, Sofia, and especially Judit, who is one of the strongest grand masters on the worldwide roster) in current competition seems to confirm that education is sufficient to generate any type of cognitive activity at highly specialized levels.

In any case, male and female brains have different characteristics, which, to a large extent, are influenced by hormonal levels, particularly *testosterone* and *progesterone*. If these differences are added to those that appear because of the different ways that children are educated in the family and in schools, the result is fundamental differences that necessarily will be reflected in the habits, tastes, predilections, and actions of each sex.

An additional problem for evaluating the differences between the sexes (the scientific name is *sexual dimorphism*) is the enormous variability among individuals of the same sex. Thus, the structural and functional characteristics of the brain vary in such a marked way among males as a group and females as a group that it is difficult to propose an average type that characterizes each sex. No conclusive studies have been carried out to determine to what extent lateralization influences men and women in their chess skills. Two recent excellent books, one by Susan Polgar (four-time women's world chess champion) and the other by Jennifer Shahade (two-time U.S. women's chess champion), explore the issue and offer numerous personal insights from the point of view of female chess professionals.

Techniques for Analyzing Brain Activity

The classic tool for evaluating the activity of the brain as various mental tasks are carried out is *electroencephalography*. One or more electrodes are placed on specific parts of the scalp to register electrical current, normally on the body of a neuron, and produce an *electroencephalogram* (*EEG*). Thanks to the EEG, numerous properties of the brain and its electrical activity have been identified in both wake and sleep states. Different types of electrical frequencies depend on the global activity of the brain. The 1 to 2 hertz (Hz) of the delta band that is identified with deep sleep is located in the lowest wavelength, and the gamma frequency of 35 to 40 Hz is identified with the wake state. This last frequency is behind the *synchrony* phenomenon that might play a fundamental role in the generation of different cognitive processes.

In recent years, sophisticated techniques of image analysis have identified areas of the brain that are active in a certain moment or as a result of a specific activity. The most important techniques are *functional magnetic resonance imaging* (*fMRI*), *positron emission tomography* (*PET*), and *magnetoencephalography* (*MEG*). The first two take advantage of the fact that active areas of the brain consume more oxygen than passive areas and therefore demand a greater blood supply. They can recognize isotopes that have been previously ingested and that are then found in the bloodstream to generate a three-dimensional map showing the regions of the brain that have more blood. The third, MEG, is a sophisticated technique that can detect in real time the magnetic fields that are created by electrical currents between neurons. The data contributed by these techniques will in the near future delimit functionally specific zones of the brain (especially in the neocortex) and describe a cerebral structure that shows that the brain acts in a collective and integrated way to carry out specific higher cognitive functions in a given situation.

Diverse studies have examined the brain's functioning during chess. This type of information is beginning to contribute important data about differences in information processing among players of different strengths. It has been discovered that during a chess game, the occipital lobe (which corresponds to visual processing) and the parietal lobe (which corresponds to attention and spatial control) are strongly activated during the cognitive processes that are involved in decision making. Other studies have verified that a player without experience shows an active hippocampus and medial temporal lobe, suggesting the analysis and processing of new information (use of short-term memory), and that an expert player shows predominantly activation of the frontal lobe, suggesting a superior order of reasoning where attention is centered in the use of already well-known mental schemas and not in trying to look for new solutions (figure 1.9). I return to this fascinating subject below.

Finally, the study of brain diseases and injuries in humans has yielded important data about the relationships between cognitive functions and the

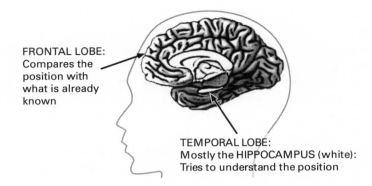

FRONTAL LOBE:
Compares the
position with
what is already
known

TEMPORAL LOBE:
Mostly the HIPPOCAMPUS (white):
Tries to understand the position

Figure 1.9
Brain activity during a chess task. Players with an Elo rating of more than 2400 use the prefrontal cortex, while novices make more use of the medial temporal lobe and the hippocampus.

regions of the brain. Think of Broca's area (located in the left frontal lobe), whose deterioration provokes the loss of the capacity to articulate language (although these patients understand what they hear), and Wernicke's area (located in the auditory cortex of the temporal lobe), whose deterioration provokes the loss of coherent speech. Numerous areas of the brain have been located and related to corporal or mental dysfunctions from this type of studies. Another valuable source of information is patients who have suffered strokes, where specific parts of the brain have been rendered useless. Thanks to these patients, it is possible to determine the kinds of cognitive processes that are associated with certain areas of the brain.

Acquisition of Functional Capacities in the Brain

The brain is not initially prepared to carry out complex cognitive functions but acquires those capacities as a result of an elaborate developmental process. Without the cultural environment in which a baby develops, neither language nor the reasoning capacity that is characteristic of humans as cultural beings will appear. The neurons' functioning is intrinsic in the sense that the structure and morphology of these cells determine their use to communicate action potentials between axons and dendrites using neurotransmitters as intermediaries. But the brain needs a full development process to configure functionally specific areas, such as the visual area of the occipital region. (It would also be possible to argue that neurons pass through a period of determination and differentiation until they reach a mature and fully functional state.) This means that, as an organ, the brain is not prepared to carry out cognitive tasks but rather needs to generate a series of relations among the neurons and with the sensations coming from its own body and from the outer world to be functionally ready.

This whole process begins very early during gestation as the brain develops. The crucial step takes place in the first weeks of development when the *neural tube* is formed from cells of the outermost layer of the embryo, called *ectodermic cells*. The closing of the neural tube marks the beginning of the development of the spinal cord as well as the brain. The *notochord*, a structure that is common to all vertebrates during development, aids to a great extent in the correct formation and later differentiation of the neural tube, as has been shown in studies carried out following specific proteins.

Once both the spinal cord and brain have closed, the true challenge of development begins. The most anterior part of the neural tube divides into five compartments as a result of a series of constrictions. The most anterior (the *telencephalon*) will become the cerebral hemispheres. It is followed by the *diencephalon, mesencephalon, metencephalon*, and *myelencephalon*, which will give rise to the rest of the structures of the adult brain. While this is happening, the neurons are starting to differentiate themselves and to emit dendrites and axons that will follow specific paths. Although dendrites establish connections only with the axons that are in their vicinity, axons will travel distances that are spectacular on the embryonic scale to innervate areas throughout the body. Axons grow to reach specific target zones by following the chemical trail left by a family of proteins called *semaphorins*, which attract and repel the axon membrane in a differential way according to their type. At the same time, many neurons die during development following a normal mechanism present in almost all tissues. Because overabundant raw material (cells) is generated, the excess then must yield its space so that the rest can grow without difficulties. This type of strategy has been taken to suggest the hypothesis of *neuronal Darwinism*, where a kind of fight for survival exists among neurons. The defender of this thesis is Gerald Edelman, who won a Nobel Prize in medicine.

Much later, each movement, each organ of the embryo, leaves its impression on the developing brain. In the human species, little by little through the nine months of gestation, the brain begins to generate a map of the world—the feet, the hands, the arms, the legs, the liver, the heart, the stomach, the muscles of the eyelids, the tongue. The whole body begins to be represented in the brain, as do certain sensations from the exterior. The mother's heartbeats are a permanent rhythm that generate a certain rhythmicity in the brain, for example, and lights coming from the outside stimulate visual responses and sounds. Everything contributes to the development of the connections among neurons, creating a map of astounding complexity that in some way is able to codify such disparate information in cellular form. The result of this cerebral representation of the body establishes proprioception, thanks to which we can determine the spatial positioning of each part of the body in an unconscious manner. Finally, for the brain to begin to carry out cognitive tasks such as perception, language, attention, or memory, it first must receive a whole series of stimuli during the first years of life. For example, visual perception is impossible if during the first months visual stimuli do not cause the cells of the retina to develop and make appropriate connections to the occipital regions of the brain.

Some Notes on the Evolution of the Brain

To analyze the evolution of the brain in primates is to analyze the origins of our species as a result of the changes that it has undergone throughout millions of years. More fascinating still is that social relations and the origin of culture and language are causal agents, found behind the evolutionary dynamics themselves affecting our species. Since evolutionary processes continue operating on any species, including humans, the effect from social relations implies that contemporary cultures throughout the planet with their different ethical and moral values exert their quota of influence on the future of the species from the anatomical and physiological point of view. But as shown above, both anatomy as well as physiology influence individual behavior, and in the final reckoning, individuals are responsible for the development of culture.

Lamentably, this circular biological-cultural relationship has been used in an abusive manner on many occasions to support totalitarian ideas and at the same time to provide a pseudoscientific basis to justify the exploitation or genocide of entire groups of humans. As an example, there are the ideas developed by Konrad Lorenz, a winner of the Nobel Prize in medicine for his work in *ethology*, the science that studies animal behavior. One of the theories proposed by Lorenz at the end of the 1930s postulated the degeneration of civilization as a result of its distance from nature and its excessive cult of urban culture. Carrying out an analogy between the phenomenon of domestication and the phenomenon of civilization, he looked for biological and evolutionary bases to justify the notion of ethnic society, without hiding his affection for the Nazi ideals that possessed, in the eyes of Lorenz, the essential characteristics for saving humanity from the degeneration into which he felt that civilization was falling. Lorenz continued to insist throughout his life on this point, consolidating himself as a defender of the ecological cause as a new strategy against urban dangers instead of promulgating, in the purest and most abominable Nazi language, the need for purity of blood in the nation-ethnic group. Evolutionary arguments lend themselves to this type of absurd analogy and the dangers that this entails. A new interpretation of Lorenz under the name of *social biology* has since the 1960s presented images of the cultural and social evolution of humans as a reflection of evolution in terms of the "fight for existence." Without losing sight of the undeniable relation between evolution and culture, I try here to keep away from this type of argument, indicating only the biological bases necessary for evolution.

Evolution is a phenomenon that operates on systems that offer fundamentally two types of characteristics—reproduction and variation. These two qualities are necessary, although not sufficient, for certain mechanisms to operate so that starting from system A one can arrive at system B. The way the dynamic works is that system A reproduces itself, giving rise to new systems that are type A but that possess a certain variation. Eventually, the variations between the original system and one of its descendants are such that a new system, of type B, is created. The phenomenon of life and biological processes on earth has generated systems that we denominate *species*. Although how to define the concept of species is an open debate, one definition is of a set of organisms that are able to reproduce among themselves. This biological definition of species means that a cat and a dog, since they cannot generate a viable descendant that at the same time could reproduce itself, are considered to belong to two different species. Species satisfy all the requirements for the evolutionary dynamic to take place: the individuals that compose a species reproduce, giving rise to more individuals of the same species.

However, reproduction generates variation in multiple ways. One way depends on genetic information, which is found encapsulated within DNA molecules. Another way of creating difference comes from the type of cellular machinery that is inherited with the *oocyte* (the maternal cell that originates the new living being when it is united to the paternal spermatozoid). It can specify different ways to carry out embryonic development, determine how the oocyte of the daughters will develop its cellular machinery, and so forth. The accumulation of variations in a gradual way, generation to generation, is called *population dynamics* or *microevolution*; the changes responsible for a species differentiating itself from its ancestral species are called *macroevolution*; and the process responsible for that change is *speciation*. In the evolutionary history of all species (including the hominid lineage that gives rise to the human species), both types of dynamics have been important.

The evolution of the brain in terrestrial vertebrates is a process that runs parallel to the evolution of other anatomical structures in the distinct lineages of this animal group. The comparative study of the brain in other animals can establish hypotheses and theories about how the brain functions that can be generalized to the human species. A large part of our knowledge about the brain is due to experiments and analysis in other mammals (mainly mice, rats, cats, dogs, and monkeys). Nevertheless, the mammal brain evolved from the brain of reptiles at some point about 250 million years ago, at the beginning of the Mesozoic era. And that brain evolved from the brain of amphibians some 350 million years ago, during the final period of the Paleozoic era. Thus, the evolutionary chain can be followed back to one-celled organisms, although with regard to brain formation, it is sufficient to begin with the chordates (figure 1.10).

Chordates (phylum *Chordata*) take their name from the presence, at least during the first stages of development, of a structure called a *notochord* that extends dorsally with respect to the longitudinal axis of the body (discussed above). This structure has a distinct construction that includes large cells of connective tissue and constitutes a kind of guide around which an organism's bilateral symmetry is organized. In vertebrates (fish, amphibians, reptiles, birds, and mammals), the notochord is lost during development, becoming part of the vertebrae. The other primary characteristic of chordates is a *nervous cord* that is dorsal to the notochord and largely formed by it. The first chordates possessed an elongated body with an anterior portion where the mouth was located and a posterior portion containing the anus. In

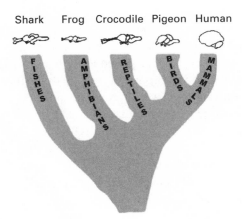

Figure 1.10
Kin relationships of the main vertebrate groups showing brain diagrams (not to scale).

this basic type of corporal architecture, nerve tissue and sense organs (such as those for hearing, sight, and taste) were accumulated in the anterior part of the body, constituting what we can consider a protobrain. The first link with a recognizable brain structure as such in the vertebrate lineage is the fish. From there, other vertebrate groups have modified or added new parts to those that already exist. Thus, the most characteristic parts of the brain, which have been described above, appear, developed to a greater or lesser extent, in the different vertebrate groups.

The area of greatest development in the brain of mammals, especially in that of primates, is the *neocortex*. In fact, the evolution of the mammal brain could be considered as the transformation and increase in complexity of the neocortex. The cerebral neocortex in humans holds three-quarters of all the synapses in the whole brain and two-thirds of the total brain mass. Here is where the associative processes and high-level cognitive functions take place. From the evolutionary point of view, the surface of the neocortex has increased considerably in the primate lineage. In addition, new areas have been added that do not correspond to areas in ancestors. New characteristics that appear are called *novelties* in evolutionary biology, and they constitute the base on which the process of change is manifested. In the aardvark (a small mammal of the African savannah), the olfactory lobe in the anterior part of the hemispheres is prominent, which has allowed this animal to adapt to situations in which the sense of smell (for enemy reconnaissance) is crucial for its survival. However, in the human species, the olfactory lobe is little developed. The same is true of the lateral optical lobe, which is present in other mammals but whose functions have been replaced in humans by the occipital cortex.

The most obvious change in the brain of the human species is its relative volume compared to the total mass of the body. The comparative study of the relative sizes of anatomical parts is called *allometry*, and thanks to this type of analysis, rules of relative proportions among different parts of the body have been discovered. These laws of proportionality are due to an increase or decrease in the speed with which a determined area develops during embryonic development in the maternal uterus (in the case of the mammals). This set of laws is called *heterochrony*, and it has been shown that the brain of *Homo sapiens* has evolved with respect to the rest of the hominid primates according to a law of heterochronic proportionality denominated *neoteny*. This law of proportionality indicates that the speed of growth of the human brain during the uterine stage (and through approximately the first year of life) is relatively much greater than that of any other primate. As a result, humans have brains that are much greater in volume than what would correspond to us by the size of our body. This has translated into an extraordinary development of the cerebral hemispheres and especially of the neocortex, with a totally unexpected secondary effect—the appearance of high-level cognitive processes.

Summary

In the structure and function of the brain, the neurons are the main protagonists of the brain's capacity to process information. Mechanisms operate inside the neurons to transmit electrical impulses and to generate memories—that is, to strengthen the relations among neurons in the synaptic connections that are made between dendrites and axons. The neocortex is the central processor for decision making, and the limbic system, especially the hippocampus, is the fundamental station for attention and working memory. All human activity has an inescapable biological base that is impossible to transcend and that influences, channels, restricts, and gives form to our cognitive capacities.

Nevertheless, within this biological straitjacket, the human brain has generated societies and civilizations based on the transmission of culture, as if this were an extension of the capacities of a brain that dreams of freeing itself from the biological yoke. Thousands of years after the appearance of the human species, as we sit down in front of a chessboard and our mental reality pauses to focus on generating thoughts based on our knowledge of the game, behind this cognitive curtain is a network of organs, cells, and molecules that allows us to construct this reality and plan our next move.

This brief explanation of the biological bases of the human mind has set out the principles and structures on which our thoughts, our critical sense, and our capacity to respond in the face of the unknown—in short, our behavior when faced with a chessboard—all rest. This concludes the first metaphor of the book—the brain as an organ whose structure allows it to construct a model of the surrounding reality. That representation constitutes a metaphor of reality that is personal and not transferable and that conditions the human mind, the subject of the next chapter.

2 The Human Mind: Metaphor of the World

Bent over the chessboard, in need of an idea, the player focuses on what move to make. He remembers that Bobby Fischer and Boris Spassky reached a similar position in their 1973 world championship match. Suddenly, he finds a sacrifice that would guarantee his victory in the battle. He has come up with a beautiful, deep move. His pulse accelerates, and his skin begins to sweat with intensity. He knows that he is playing the last game of the match and that if he wins, he takes first prize. He looks around and remembers that he is in Madrid, where he arrived two weeks ago to play in the tournament. He returns to the board. There are only three possible variations, and all three seem to give him the win. He goes over them one by one and confirms with satisfaction that the result would be favorable. Suddenly, he stops. A look of annoyance shows on his face. His muscles tense, and he begins to jiggle his legs. He has found a rebuttal to his sacrifice in the third move of the second variation. The sacrifice won't work. His body relaxes, and he starts to look for another idea.

The brain operates in several ways during this episode. Perception as an act of sensation and of understanding makes the pieces on the board a series of symbols, infusing them and their positions with concrete meanings in the mind of the experienced player. Long-term memory is in charge of validating present perceptions in accordance with previous knowledge: the player remembers the position that he is perceiving and looks for plans based on what he has already experienced. The discovery of a move—here, a sacrifice—generates a series of emotional responses that are transmitted to the body in a feedback relationship. The search for variations puts a logical thought module into play. After finding a rebuttal to his idea, the player experiences sensations of frustration, a new emotional response that is transmitted by the body. Perception, decision making, logical thought, memory, emotions: these are some of the brain's emergent functions that make up what we know as the mind. Joined with those functions are the player's perceptions of himself, of his body, of spatial consciousness, and of being in a specific place in the world that is full of meanings that run across his brain. When he realizes that he is in Madrid, the scent of patatas bravas (a dish made of potatoes with a spicy sauce) calls forth memories and sensations that unconsciously influence the decision that he will make for his next move.

Mind-Brain Duality

In the previous chapter, the structure and the most important functions of the brain were discussed. The problem to consider now is the following. When analyzing the behavior of most animals other than our own species, it is apparently clear that an animal's ability to relate to the environment can be described as a function of its brain's capabilities. Thus, we do not have any problem in assigning to the brain all the functions that are necessary for the survival of an organism and even for carrying out mental processes such as learning. The trained dog gives the ball back, the carrier pigeon reaches its destination, and the dolphin makes acrobatic leaps. Nevertheless, when trying to understand human behavior, we resort to the existence of some-thing different or something extraordinarily different. In fact, this classic problem of Western philosophy has been transferred, like so many others, to the realm of science. Sometimes, the transposition from philosophy to science takes place in a clear and precise way, as is the case with the origin of the universe or the presence of life on other planets. Both problems passed from the metaphysical domain to the scientific domain without the need to reconsider the nature of the initial question itself. In the case of the origin of the universe, science tries to decipher the available clues (such as traces of residual radiation or spectrographic analysis of the light radiated by stars), generating hypotheses that predict a series of phenomena that can be veri-fied by observation and experimentation.

However, other problems (such as the nature of life or a definition of what the concept of life means) have suffered much more reexamination in moving from metaphysical speculation to the scrutiny of science. In cases like these, the fundamental problem is to reach a workable definition of the concept. The issue is defining what we understand by life, for example, and seeing how its characteristics and properties can be explained in a rational way.

The same thing happens with the problem of the mind. Science has inherited a poorly structured problem from the multiple philosophical speculations that have focused on it—dualism, materialism, monism, epi-phenomenology, connectionism, and functionalism, to mention a few. So to advance in understanding the phenomena that accompany the concept of mind, scientists must undo many of the metaphysical preconceptions that have constituted an enormous intellectual force for more than two thousand years. These metaphysical ideas have been systematically nourished by a religious worldview that is still ingrained, even in science.

Consequently, it is almost impossible to approach a scientific problem without prior metaphysical prejudices. Contemporary philosophy has tried to recycle itself by using the information that comes from scientific disciplines (such as neurology, psychology, psychiatry, biology, ethology, and cybernetics) that contribute to the elaboration of theories about animal behavior. Today, philosophy of the mind is contributing to the clarification of a fundamental problem in our conception of the nature of the human species and its position in the world. This metaphysical prejudice is the so-called dualist conception of the nature of man that was inherited mainly from Plato but finds its clearest exponent in the works of French philosopher René Descartes (1596–1650).

Descartes is credited with the idea that living organisms are like machines, furnished with mechanisms that humans could build. It is an extremely simplistic metaphor (see the following chapter), but at the same time it recognizes the phenomenon of life as an integral part of nature, offering a metaphor with which we can begin to understand the nature of life itself. In that sense, we are indebted to Descartes for taking a scientific attitude in the face of the processes that characterize our own nature. Nevertheless, through his famous dictum "*Cogito, ergo sum*" (I think, therefore I am), Descartes insisted on offering a proof of the existence of a duality between body and soul. For Descartes, the body is a machine that, when faced with external stimuli, reacts through reflexes that are controlled by an immaterial entity that is devoid of any dimension in space, denominated soul, or mind. Thus, the brain, as a part of the body, stays separated from the mind, which is outside the reach of human understanding. But even in this vision of nature where the mind is separated from the body, it is necessary to explain why a physical event such as a blow causes a mental sensation like pain or how the mental decision to write a book is translated into a series of physical acts to carry out that decision. Descartes postulated the existence of a communication channel between soul and body (between mind and brain) that resides in the pineal gland, a gland chosen, apparently, solely due to its central position within the brain. Nowadays, the pineal gland is known to secrete melatonin, a compound that is involved in different types of regulation, from circadian rhythm to the regulation of sexual appetite.

Emergent Complexity

This dualist view, which posits a mind-body separation, has been difficult to cast off. In philosophy of the mind, epiphenomenalism postulates a dualism that is more in tune with scientific explanation, since the organism is understood to be a material entity and the mind an *epiphenomenon*, or a phenomenon that emerges from brain activity. In any case, in a world that is strange and filled with inexplicable events, the search for complexity has been burdened with many biases that are not easy to shed. The complexity that accompanies cognitive processes (such as the generation of thoughts and feelings, the elaboration of moral and ethical principles, and the development of art or of science itself) would seem to indicate that a quality exists in human beings that distinguishes and separates them from the rest of the living beings that make up the scale of life. And in that Aristotelian *scale of nature*, humans see themselves as the superior link. This search for metaphysical meaning in human existence, where the human species is granted a privileged position in nature, is a refined version of dogma and religious prejudice. What makes us more complex than any other organism? Why do we insist on looking for and explaining that complexity? Is it real complexity or simply a reflection of the search for a reason that convinces us of our superiority?

The idea that the mind is the place where the key to our superiority resides has attracted the imaginations of a great number of researchers, artists, and charlatans who have searched for simple and complex answers. Consider the following. Four billion years ago, the only living beings that inhabited the earth were the prokaryotes, organisms that lack a nucleus (a separate membrane protecting genetic material) and whose best known representatives are bacteria. These organisms reproduce at such vertiginous speeds that in a few hours, a colonizing bacterium can give rise to millions of descendants. These wonders of nature are also able to adapt to any type of atmosphere—from the marine depths to the rim of a volcano, in the presence of oxygen or its absence, within a mammal or on a book. Given these circumstances, it seems somewhat difficult to affirm that a multicelled organism (such as humans, for example) that appeared on the scene of life no more than a million years ago is superior to the bacteria. Nevertheless, humans carry out activities that not even the most recondite, anaerobic, promiscuous, thermofilic bacterium could imagine, especially because it lacks imagination. My friend, the Spanish scientist Jorge Wagensberg, expresses the idea in the following way: "Between a bacterium and Shakespeare, something has happened." There is an answer to the question of which species is more complex, but not one that favors us exclusively. Certainly, a bacterium would have a tough time trying to write *The Tempest*, but by the same token Shakespeare would not last more than a few seconds in the abyssal zones of the ocean. They are two different complexities that have solved the enigma of existence in two extremely different ways, but both of them are complexities after all.

Nevertheless, it is true: something happened. Life without consciousness of itself, even though it can develop under highly adverse conditions, seems to correspond to a level of lesser complexity. So what has happened? The answer surely resides in the development of the brain and its capacities as an organ that is specialized in generating cognitive processes. But there are levels of complexity in different groups of organisms' relations to the environment. Bacteria have one degree of complexity, plants another, worms yet another, lions, dogs, cats, cows: where do we stop? When we reach the group that humans belong to, the primates, the differences begin to narrow and, in many cases, to disappear. However, Wagensberg's statement continues to be valid whether we are talking about bacteria or chimpanzees: only our species is able to generate a work of art. This is the moment to begin to investigate the processes that occur within the brain that make that possible. When we do so, we will discover that the mind is nothing but the integration of cognitive processes and that these in turn correlate with different states of the brain that are determined by neuronal activity. The philosophical position that the scientific community takes with respect to the mind is, therefore, monist: the brain generates the mind. Nevertheless, the problem of identifying the mind as a state of the brain is not so simple. The first obstacle is that of consciousness or, to be more precise, the diverse conscious states. More complex still is the fact that the mind elaborates images and thoughts in response to stimuli from its surroundings and then also is conscious of itself—that is, in addition to generating a representation of the world, it elaborates a representation of itself. To complete the complications, the human being, as a social animal, elaborates hypotheses about the minds of other human beings. Let us try a simple experiment. Close your eyes for a moment, and try to create an image of this book. How many other thoughts and images pop up along the way?

Three Levels of Abstraction and Two Operational Spaces

The integration of the cognitive processes as a result of the coordinated action of different areas of the brain forms what is commonly known as the human mind. Following the traditional division made by German philosopher Immanuel Kant (1724–1804), cognitive processes can be separated into three large groups that appear, hierarchically organized, in a variety of organisms. These large families are the processes of sensibility, understanding, and reasoning. Neurobiologists Stanislas Dehaene, Michel Kerszberg, and Jean-Pierre Changeux consider that these three levels represent three levels of abstraction (more on this below). Thus, the *sensibility* processes form the first level, creating representations of the world using inputs from the sensory organs. The *understanding* processes then in some way organize the information coming from the lower level of the hierarchy, generating concepts. Finally, synthesis of the concepts is carried out by means of the *reasoning* processes, which constitute the last and most sophisticated level of the hierarchy. Within each of these levels, we can distinguish a series of specific processes, some of them very familiar, which through their integration form the real base on which the processes that give rise to the mind (consciousness and intelligence) rest. They are, of course, the same processes that are necessary to carry out the complex cognitive activity of the game of chess—proprioception, perception, memory, learning, thought, attention, problem solving, decision making, creativity, and feelings. It is often impossible to delimit a clear border among the different processes and to separate them out from the extraordinary integration that is generated in a given moment, causing what is called a *mental state*. In the same way, to complicate the problem even more, those processes that we share with the majority of animals that possess a nervous system (hunger, thirst, fear, and so on) are continually acting on the periphery of the cognitive processes.

Numerous explanations have been proposed about how the brain is organized to generate cognitive processes. The main problem is to combine our knowledge of brain anatomy and function with cognitive capacities. One problem is to identify the specific structure (for example, the hippocampus), another problem is to determine what type of function is carried out from the point of view of the transmission of electrical impulses (where they are received from and transmitted to), and another very different problem is to determine what specific type of cognitive capacity is involved in this structure. These problems can be studied using diverse types of analytical strategies. Thus, studying anatomical structure requires histological preparations, analysis of neuron morphology, and analysis of the types of connection patterns that exist. Studying current transmission requires methods like those that were discussed in the previous chapter (such as EEG, MEG, PET, and fMRI). Finally, determining the type of cognitive activity in which the cerebral structure in question is engaged requires experiments with live animals (including humans) during which the structure's activity during a specific cognitive effort can be revealed. Besides these experimental strategies, studying the decline in cognitive capacities in individuals who have undergone a stroke or who have had specific parts of the brain removed to alleviate diseases such as epilepsy or tumors reveals important information about the way that the brain operates to generate mental capacities.

The Dehaene, Kerszberg, and Changeux model explains the organization of the brain in relation to cognitive processes, integrating the empirical data coming from different research fields. Their model postulates the existence of two operational spaces in the brain—a global workspace and a set of modules that are charged with carrying out specific cognitive operations (figure 2.1).

These modules operate like processors and are specialized in the general cognitive tasks of perception, motor systems, memory, evaluation, and attention. In humans, the perception module would be located in the lateral and ventral areas of the temporal lobes and in the lateral and inferior parietals, including Wernicke's area. This module allows the content of any object or perceived event to be accessed in the global space. In other words, it allows access to what is happening in the present. The motor systems module would include the premotor cortex, the posterior parietal cortex, the supplementary motor area, the basal ganglia (especially the caudate nucleus), the cerebellum, and the language-production areas of the inferior left frontal

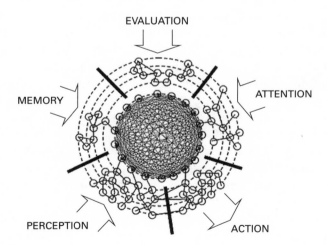

Figure 2.1
The Dehaene, Kerszberg, and Changeux model of how the brain generates cognitive processes. The operative global workspace (in the center) coordinates cognitive modules, putting together the past (long-term memory), the present (perceptions of what is going on), and the future (ideas of actions that might be taken) jointly with attention (short-term memory) and evaluation.

lobe (which includes Broca's area). Thanks to this module, the global space generates a motor or language response as a result of the representation that is found in those moments in the work space—that is, it guides behavior and future intentions. The long-term memory module, distributed throughout the cortex according to memories' content and the way they were stored, would be mediated by the areas of the hippocampus and parahippocampus. This module provides the global space with access to memories, or in other words, it allows access to the experiences of the past. The evaluation module would be located in the orbitofrontal cortex, the anterior cingulate, the hypothalamus, the amygdala, the striated ventral cortex, and the mesocortical projections from the prefrontal cortex. Thanks to this module, positive or negative emotions are generated about the global space's current representation—that is, it maintains the system of values and affection. Finally, the attention module would be specially located in specific areas of the parietal lobe involved in visual and spatial attention and also in the rest of the descendent projections from the global space to other modules. Its function is to selectively amplify or attenuate the signals coming from the rest of the modules—in other words, to allow concentration on a specific event or memory.

The global space of operations would be determined by a set of neurons that are distributed along the width of the neocortex and are characterized by their ability to receive and transmit information to other areas of the brain through horizontal projections with long axons. These pyramid-shaped neurons probably have their origin in layers two and three of the cortex. Besides the horizontal projections, the neurons have vertical connections toward the interior of the brain through layer five. The global workspace neurons would be mobilized to carry out cognitive tasks for which use of the specialized modules is not sufficient. In this way, the global space is able to invent new abilities simply by using variable proportions of each module.

The dynamics of how this model functions predict the spontaneous action of specific subgroups of global space neurons depending on the particular state of the brain, which results from the representation created by the memory or perception module. Thus, the capacity for generating diverse states is very large, granting a great amount of flexibility to the brain's functioning and allowing it to carry out a multitude of cognitive tasks. Control and modulation systems would exist for the global space neurons (such as the action of the reticular neurons, for example) that control the transition between the wake state and non-REM sleep. At the same time, other types of modulators would have to exist to increase or inhibit the degree of activation of the global workspace. For example, activation would be increased when faced by an unexpected, novel fact or when there are very strong emotions, while activation would be reduced when carrying out routine activities.

Moving on from this plausible model of the connection between mind and brain, I will now cover some of the most relevant cognitive capacities, which will later help us analyze the mind of a chess player.

Proprioception

Proprioception is the sense of individuality. This mechanism constantly maintains a clear demarcation between the organism and its surroundings. It is mostly not a conscious process. If the mind were occupied in identifying each muscle of the body and in processing this information to know the muscles' condition and location, it would find it impossible to generate any type of activity. Proprioception begins during the development of the embryo and the fetus as brain representations of each muscle, integrating in the first years of life to give rise to the sensation of knowing oneself as an individual in a given position and point in space. The brain regions involved in proprioception are the most primitive from an evolutionary point of view. Little by little throughout embryonic development, the brain generates the first representations of the world and of the individual itself. It is probable that embryonic movements provoke an assimilation response for a given series of motor neurons. The muscular fibers contract spontaneously, which stimulates the nerve fibers to innervate them, which in turn stimulates the brain. The same thing must happen with the internal organs. As a result, the brain begins to have a map of the body in the form of multiple networks of synaptic connections. Gradually the fetus begins to have a representation of itself, in which the neurons distribute the information that comes from the body. Body and brain begin to be—are—indivisible. When the baby is born, the brain continues to represent the world (the individual's internal world and the world that surrounds it) as a result of the stimuli received. Stimulation is fundamental. For example, experiments with mice have demonstrated that a juvenile that grows up in the dark does not develop the capacity to stimulate the cones and rods of the retina and thus is virtually unable to develop sight.

Perception and Knowledge

Perception consists of a series of processes through which we receive information from the outer world. The sense organs (sight, hearing, touch, smell, and taste) are what put us in contact with a changing world. Through the perception of these dimensions of reality, we create for ourselves an idea of the world that surrounds us. The brain integrates the information coming from our senses and generates mental states accordingly. For example, cold is a state of the mind in which the individual is conscious of a corporal state that is characterized by a surrounding temperature that is noticeably inferior to that of its body. These kinds of sensations (cold, heat, sleepiness, thirst, hunger) directly relate our mind to our brain. The nerve endings that are sensitive to low temperatures generate a mental state that is integrated within the brain, causing the individual to understand what state it is in. In humans, this understanding goes hand in hand with language characterization, generating a thought such as, "How cold it is!" With perception, knowledge of the world begins, and through this knowledge comes access to levels of abstraction that generate ideas about the world that do not come directly from the senses. Here we are faced with the fundamental problem of knowledge and its sources. Where does knowledge come from? The simple answer grants a fundamental role to perception. Nevertheless, a philosophical current since Kant postulates the nativism hypothesis, arguing that the brain has predispositions (spatial sense, temporality, and causality) that allow us to mold or give meaning to what we perceive.

Pattern perception (that is, the perception of similarities in spatial or temporal configurations) has a fundamental role in playing chess (as is discussed further in the next chapter). An enormous amount of research has been carried out about pattern recognition, and it continues to be a very fruitful research field. This kind of perception is an active perception in the sense that it is accompanied by understanding. There is also perception without understanding. For example, when we see something for the first time, the mind perceives an object without being able to assign a meaning to it no matter how much it is put into context, unless there are sufficient clues to deduce its nature.

Four levels of morphological organization can help us with the representation and analysis of an image—proportion, orientation, connections, and articulations. The first three are static qualities, while the last is a dynamic quality. To these, we can unite qualities that are not strictly morphological, such as color, contrast, and brightness. What happens when we look at a chessboard (figure 2.2)? At first sight, the board with its sixty-four squares of alternating colors delimits a space. This space is loaded with meaning for the expert player, including the center (squares e4, e5, d4, and d5), the queen-side and kingside, the first rank, the last rank, the files, and squares f2 and f7. These areas convey information all by themselves, without needing to be occupied by pieces.

Connectivity plays an important role here, since the squares have adjacency relations among themselves. Each square is surrounded by eight others,

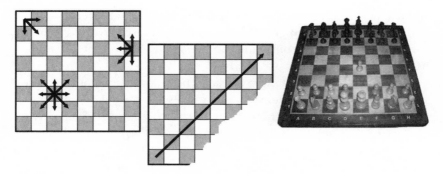

Figure 2.2
Some perceptual elements on a chess board. *Left:* Adjacency relationships specify the number of immediately neighboring squares (three on the corner, five on the edge, and eight in the middle). *Center:* Contrast between colors forms the diagonals. *Right:* A real board is worth a thousand diagrams.

except those on the edges (which have five neighboring squares) and those in the corners (which have only three). Color also helps with recognition of adjacency relationships, especially in identifying the diagonals. Finally, the proportions of the pieces are also fundamental for recognizing positions. Because of this, chess masters demand adequate lighting in major competitions, and more than a century ago, a standard type of piece (called Staunton) was adopted for use in matches. No professional chess player would agree to play a match with crystal or fantasy-themed pieces. Changing all the perceptual content that a chess master is used to interacting with would surely change the way that he or she acts. Nowadays, thanks to computer chess programs and real-time Internet game software (see chapter 5 and the appendix), more and more players are learning to perceive the board and pieces in two dimensions, drastically changing the type of visual experience that is involved and the quality of the cognitive processes that are activated when playing. These differences have not been studied in depth, but the chess scene is radically changing. It would be of great interest to evaluate these Internet-based differences in a controlled context.

Memory and Learning

The human brain, like the brain of most animals, is capable of learning from past experiences. It therefore needs to be able to codify the information coming from the exterior world in a way that allows this information to be stored and retained. At the same time, the brain needs a system to access what has been stored (remembered). A system with *memory* capacity requires four fundamental properties—codification, storage, retention, and information retrieval.

Additionally, it is important to consider that information does not necessarily come from the exterior world but can also be created internally from thought. In other words, a thought can be coded and stored in memory for later retrieval. When the mind sails past the complications of a chess position, it invents variations that exist only in the form of synaptic connections within our storehouse of immediate memory, the *hippocampus*. Expert players remember those variations as if they were seeing them—as if they were altered forms of the patterns that are stored in long-term memory throughout their professional experience and distributed in the cerebral cortex. Nonexpert players, on the contrary, see different variations as ephemeral—as disappearing when the game is over or even in the very moment the player is thinking about what move to make. Sometimes short-term memory plays tricks on us during a chess game, with the player confusing one variation for another and the result ending in disaster. Nikolai Krogius proposes an example of what he calls *residual image* from an Ilin Shenevski-Vladimir Nenarokov match (Moscow, 1923). The player forgets that the bishop is unprotected. This basic error can be avoided simply by looking at the board before making the move instead of getting carried away by the (incorrectly) remembered variation (figure 2.3).

Memory is closely related to learning. One of the key aspects of learning is the retention in memory of either a fact or a series of clues that leads us to deduce a fact. In the first case, we can memorize a friend's telephone number, and in that way we have learned how to communicate with him. In the second case, we can memorize the geographic situation of a river on a map because we know that such and such town is located on the river, and so we have learned a key to finding the town on the map instead of memorizing (learning) its exact location. In chess, this distinction is fundamental: it is the difference between learning an opening's moves in a mechanical way and learning the position you reach by means of the opening. One of the characteristics that distinguishes us from the vast majority of other animals is that human beings are able to transmit something learned to another person. This means that it is not necessary to have a concrete experience of a certain fact or phenomenon to learn it (which brings to mind police witnesses who swear they have seen facts that in reality have been suggested by others).

This *cultural transmission* exists, although in a more precarious way, in other animals. One member of a population of Japanese Bonobo monkeys,

BLUNDERS

Figure 2.3

Example of an error in short-term memory. White wants to open the f file, not realizing that his bishop is unprotected. The variation played speaks for itself: (1) ♖f1 g6, (2) ♕e3 ♘e7, (3) f4?? exf4, attacking the bishop and the knight at the same time: (4) ♕xf4 ♕xb5.

for example, discovered how to use water to wash sand off a potato before eating it, and this knowledge spread throughout all the population. In other words, there was a process of cultural learning. There is no doubt that in the evolutionary past of the human species, this capacity for collective learning is and has been responsible for the creation of social networks, which has brought us to the formation of what we now call *culture*.

The brain's storage systems, commonly known as *memory*, allow the learning process to be carried out in a way that the concepts of memory and learning can almost be considered equivalent. In this sense, to learn is to retain in the memory facts from the past (or, in Platonic doctrine, to recall eternal and immanent ideas to our own biological structure). But to learn also has a clear cultural connotation: what has been remembered must be useful as a source of information for our future actions. That idea is the source of sayings such as, "Those who cannot remember the past are condemned to repeat it." There is no experience more frustrating than to recognize a board position but not know what it means—to remember only the position of the pieces (because you've played it in the past) but not its meaning or the way to continue.

Little is known about the biological process by which an experience is stored in biochemical form in the neural network that forms the brain. (The use of NMDA receptors, discussed in the previous chapter, seems to be one of the cellular mechanisms that are involved in memory.) The synapses are known to be involved with this information transmission—at the experience level (as a means of interchange between the person and his environment), at the symbolic level (as a means of understanding the experience), and at the subsymbolic level (defined by biochemical processes). It is easy to think about the recording of visuospatial experiences that could remain in the brain as connective replicas of a visual impression (that is, like an isomorphic model of the visual impression itself), as if it were a map that the brain draws up of the image in the form of neuron connections. But how is it possible that the scent of my grandparents' room has been recorded in my brain in the form of synapses? There is no direct representation of smell-synaptic connection, as with space-synaptic connection. Apparently, memory experts think that all our records are stored in the form of proteins within neurons. Nevertheless, that is far from knowing how a molecular accumulation is equivalent to a symbolic concept like love or understanding how a coder/decoder translates from one to the other when we learn something or remember a fact already learned.

The representation of the world within the brain is a process of biochemical cellular codification that begins during intrauterine development. When we begin to use and learn complex cognitive activities (such as language), the brain carries out this codification automatically, taking as its departure point the codification from primary sensory stimuli. This means that high-level cognitive activities (like thought, imagination, and logic) are more sophisticated and more abstract than the rest of the cognitive functions. And the fact that the brain can codify in the form of multiple synapses an abstract concept like love is as simple at a cellular level as codifying the sight of an apple. All that is needed is for the brain to undergo an adequate development and learning process during which the neurons become coder/decoder machines thanks to the specific proteins that are being transcribed and the synaptic connections that are being generated.

In this way, the brain represents the world—beginning with the cellular codification of its own body, followed by sensory stimuli, and finally, as a result of learning during childhood, the codification of another type of stimuli (such as language). Categorizing stimuli is a fundamental process in this sequence of representation of the world, and it is done in a way that allows the mind to distinguish similarities from variations. Similar stimuli find easier access to the memory and therefore can be learned more easily than those with novel dimensions. For this reason, an expert can more easily remember facts and events related to his field, given that he can relate them to others of the same type that have already been stored and organized in his memory. As is discussed in chapter 3, strong chess players are able to remember positions on the board with much more facility than weaker players can.

Memory as a mental process is divided into different systems with distinct properties, distinct ways of operating (in terms of storage and retrieval), and possibly distinct ways of codifying protein and synaptic forms. There are up to five types of memory with diverse subtypes, depending on who is doing the classification. The two most important are *short-term memory* (which includes working memory—what we use to carry out daily activities without losing track of what we're doing) and *long-term memory* (with its two major and important subdivisions—*episodic memory* and *semantic memory*) (figure 2.4).

Episodic memory stores facts and events as part of the personal history of an individual, while semantic memory stores facts about the world (that is, it forms the warehouse of our ideas or representations of the world). To these two types of long-term memory are added *procedural memory* (which stores habits and abilities) and *perceptual memory* (which allows us to remember

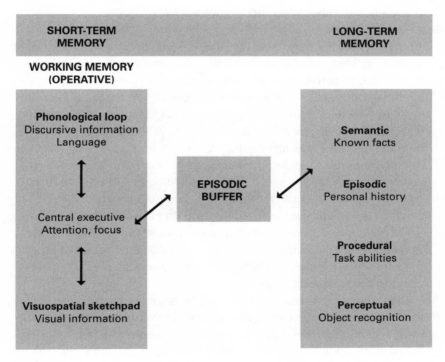

Figure 2.4
Types of short-term and long-term memory.

and identify an object after a stimulus). As has already been indicated, *working memory* allows an action of greater or lesser complexity to be carried out without losing track of what is occurring. This memory adds a cognitive component—information processing—to short-term memory, bringing it close to being a measurement of an individual's intellectual capacity. (This point is looked at again in the discussion of the concept of intelligence.) Thanks to the work of Alan Baddeley, we can distinguish three centers within working memory—the *central executive*, the *phonological loop*, and the *visuospatial sketchpad*. The central executive is the most important part of the system, and the other two depend on it. As Baddeley indicates, it is a system that allows cognitive tasks like playing chess to be carried out. Besides these three centers, an episodic buffer allows what is happening to be stored before sending it to long-term memory. While the central executive bears the load of attention to what is happening, the loop and the sketchpad take account of the discursive information coming from language and visual images, respectively.

The Magical Number Seven (Plus or Minus Two)

A fundamental discovery about how memory functions that concerns learning is the existence of a limit to our capacity to retain events in working memory. In an influential article in 1956, George Miller proposed the existence of a "magical number" for the maximum capacity to retain information—seven items (plus or minus two). A key aspect of Miller's model is that these items are highly organized and depend on our past experience. The process of reorganizing the information that we perceive is called *recoding*. The items or groupings of information that are organized in a way that makes them meaningful are called *chunks*, or information modules. As is shown in chapter 4, an expert player can remember a position on a board just by looking at it for a few seconds, since he groups the pieces in modules that are loaded with relational information. If he had to remember the position of twenty isolated pieces or ones placed at random on the board, the lack of meaning in their mutual relations would make the task impossible, as in fact has been demonstrated in experiments.

The theory of information processing by means of modules can explain the fact that we remember a great number of events in working memory using regrouping tricks that form a kind of immediate memory hierarchy. When we try to remember a telephone number, we can remember the first three numbers and know that they correspond to a specific area of the city. So a nine-digit number like 914168268 can be grouped as 91 (Madrid's area code) plus 416 (the Prosperidad district) plus 8268 (the four numbers that must be remembered). Nine numbers have been reduced to four. For anyone not familiar with the telephone numbers of Madrid or Spain, the task of memorizing the nine numbers is more complex.

For a chess player who is familiar with the fianchettoed bishop, remembering a complex position in a chunk with pawns in f7, g6, and h7, king in g8, rook in f8, knight in f6, and bishop in g7 is as simple as remembering the position of a single piece. Moreover, the fianchetto formation carries with it numerous other connotations—such as domination of the long diagonal, an empty h6 square, and many other dynamic ideas that are known only with experience and learning (figure 2.5). Professional chess players have many of these chunks of pieces and squares full of significance that turn the board into a meaningful universe and that let them recall the details of a position with a simple glance. In the eyes of a less skilled player, this memory seems incredible, but for the master player it is simply a trivial matter.

All types of memory are important for carrying on the normal activities of daily life. For example, without episodic memory, we would not have any notion about ourselves. We would not know what happened to us

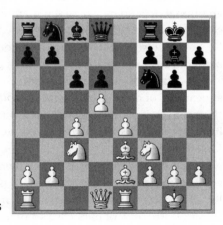

PERCEPTUAL CHUNKS

Figure 2.5
Position after move 8 by white (Karpov-Kasparov, 1990, World Championship). Black has a fianchettoed bishop in the middle of a complex position with all thirty-two pieces on the board. The twelve highlighted squares on the black kingside form a structure full of meanings. Any professional player (and even a typical club player) could quickly memorize the content of these occupied and empty twelve squares. A professional player would add a semantic content to the position. She would instantly know that it is the classic variation of the KID (king's Indian defense, E92 in ECO code) and could show the order of the moves that originated the position in the diagram.

yesterday, when or where we went to school, or who our parents are. Without short-term memory, it would be impossible to carry out a task as simple as calling someone on the telephone. After lifting the receiver and dialing the number, we would have forgotten whom we wanted to call. And without semantic memory, we would not even know what a telephone is for or how it works. In chess, each type of memory has its importance. Semantic memory lets us maintain a repertoire of knowledge about general principles, both strategic as well as tactical, as well as a memory of games or positions that have been studied. Episodic memory maintains a registry of our experience from games that have already been played. And working memory allows us to understand the position and generate a game plan. (These aspects of the brain and chess are discussed in chapter 5.)

Emotions are also related to memory. It is easier to remember an event if the emotional state that we had when it was memorized was consistent with the emotional load of the memory. People who suffer from depression, for example, remember sad events much better than happy ones. It seems likely that actors who simulate the mood of the role they are playing while learning dialogue would remember their lines better during a performance. There is also some evidence, although apparently less conclusive, that there is an emotional dependency between memory and an event. In these cases, the context from the emotional load is used as a memory clue. For example, if we learn a lesson while we are in a good mood, it will be easier to remember it during an exam if we are again in good spirits. Emotions seem to play an important role in the codification and retrieval of memories. Finally, the appearance of *traumas* (*traumatic memories*) and the repression of memories that were codified and stored in stressful situations are of interest to fields like psychoanalysis, which centers on phenomena such as intentional forgetting.

Forgetfulness is another aspect of memory. Indeed, it is not possible to forget what has not been memorized. As the Spanish psychologist José María Ruiz Vargas comments, forgetfulness might help with remembering new things. If we think of memory as a process where the codifying, storing, retaining, and retrieving of information take place, then all sensory impressions are codified when they reach the brain, but the amount of storage and retention depends on both the intensity of the stimulus and the context in which this stimulus is received (which is related to the intensity). Information that is not rehearsed is lost for good and not stored in short-term memory, and so it cannot pass to long-term memory. Nevertheless, there are many cases in which you remember events that you have not paid attention to, although normally these are contextually associated with something that has been processed by our attention. To differentiate these unconscious processes of memory acquisition, they are called *unconscious* or *implicit memory* (versus *conscious* or *explicit memory*, which is a memory created by paying attention). Since consciousness is a novel process in animal evolution, implicit memory and learning must evolutionarily precede explicit memory and learning and therefore must be much more generalized than the latter. Most of our memories come from an unconscious process, and all our initial learning during infant development precedes conscious capacity.

An interesting fact about memory and forgetfulness is the relation that these have with dreams and REM (rapid eye movement) sleep. Indeed, diverse theories have been suggested about the functional value of dreams, although no consensus has been reached. Whereas thinkers like Sigmund Freud assigned dreams the function of a window to the unconscious, other researchers have considered dreams merely a secondary effect of neural activity. Nevertheless, dreams can serve (which does not mean that they are designed for it) as a consolidator of images, memories, and learned events, as demonstrated by the fact that after a learning task, individuals who, under experimental conditions, are allowed to enter REM sleep remember much better what they have learned than those not allowed to do so. A rich literature exists (led by work from Allan Hobson's group at Harvard University) about dreams as a conscious state of the mind and a brain activity that is generating new and important ideas in relation to the problem of consciousness.

Memory and learning are tightly linked, and in many senses, they can be used synonymously, since to learn something is to memorize it. Consequently, a theory about the acquisition of knowledge must be based on a theory of memory. The more we advance and deepen our understanding of memory systems at all levels (from the molecular to the cultural and including the cellular level of synapses and neural networks), the closer we will be to having a complete theory about how the human species acquires knowledge.

There are different types and strategies of learning. On the one hand, there is *classical conditioning*, studied by the Russian physiologist Ivan Pavlov. This mechanism of learning is based on the fact that by associating a stimulus to a response, an animal can be conditioned in a way that a stimulus that is not directly related to a response can still end up triggering it thanks to an association with an *intermediary stimulus*. The classic Pavlovian example is the dog that is trained to associate the presence of food with the sound of a bell. Eventually, the sound of the bell stimulates the secretion of saliva because it is directly associated with food, even if food does not appear. Other types of learning that are not based directly on association (although many researchers would reduce them to a series of associations) are *habituation* (an action is repeated until it is learned without any type of reward), *sensitization* (a neutral stimulus sensitizes a response), *trial and error, latent learning* (an action is learned without any motive and later helps someone learn to solve concrete problems), *imitation*, and learning through *hitting on a lucky solution* (a new, complex problem is solved by a fortunate working out of relations among known elements).

Attention

Attention is an essential element of the brain's activity. If we could not concentrate our mental activity on a specific problem, we would be incapable of carrying out any conscious action. Instead, we would be occupied with stored images and memories. We have seen that attention is closely related to working memory. In chess, the discovery of a salient characteristic of the position demands attention, and the rest of the board is relegated to the background. This type of discrimination is critical to differentiating between an expert player and a nonexpert. The capacity for concentration and, in particular, the capacity to pay attention to what is important are fundamental attributes of experts in any field of knowledge.

Thought

There is no operative way to define clearly what it means to think. Up to a certain point, thought is the internal elaboration of a representation of the world. It is a brain activity that, instead of producing a motor response (an action), produces a kind of internal image. This inner image is private (we do not share it with anybody), and sometimes (as certain mental disorders such as schizophrenia demonstrate) it can be as vivid as a representation of reality. A patient with schizophrenia hears voices in his mind that he identifies as coming from the outside world (many visionaries who assert they speak with divine entities suffer from schizophrenia). Therefore, one characteristic to describe thought is its internal nature. Instead of provoking a motor response, it provokes a new association in the mind.

Thought and action are so closely united that one of the classic theories about the nature of thought considers it to be a motor action like any other. This theory says that we generate movements while thinking. Thus, small movements of the tongue are especially significant when you are talking to yourself in your thoughts. Experiments where muscular movements are totally suppressed by the controlled administration of curare (a mixture of alkaloids from plants of the Amazon that indigenous peoples place on the ends of darts or arrows to immobilize their prey) have demonstrated that thought follows its course in the absence of a motor action. In any case, when we have a thought, we often generate a movement. Consider the typical movements that take place while thinking about a problem or a situation, such as biting the lower lip, pursing the lips, wrinkling the forehead, scratching the forehead or chin, and rocking. Children, who have still not learned to control their body movements, can be observed moving and contorting their bodies when trying to solve a complex problem. These movements are considered a type of excess motor activity that is produced by the frustration of not being able to solve the problem. As children grow, they learn to control those movements, which were nothing but the motor manifestation of thought.

Other theories assign to either images or language the key role of being the essential element of thought. In the first case, thinking would mean elaborating an image with the mind; in the second, it would be putting together a phrase. The fact that many people can elaborate thoughts without resorting to any image and that others can elaborate vivid images without needing to accompany them with phrases indicates that neither one is the essential component of thought. Neuroscience still does not have a precise idea of what constitutes thought at the level of mental process. What is known is that one thought links to another and that, in the absence of an objective or a direct stimulus, continues provoking associations until some type of action accompanies the thought. At that moment, the thought becomes a prelude to decision making. In this way, thought and decision making are intimately linked. Even when we are wandering in our thoughts without any apparent fixed object, letting the mind create this or that image, a series of associations is generated that culminates in a decision such as, "Now I'll get up and make a cup of tea."

Decision Making and Problem Solving

Most situations in daily life generate the need to elaborate an answer (unless you are a Taoist monk committed to nonaction). As discussed above, the simplest type of answer is the reflexive response, such as when we immediately pull our hand away from a hot plate. In a multitude of other situations, the mind is fully conscious, and a response requires more complex elaboration— that is, it is necessary to make a decision. For decision making, various cognitive properties are needed (such as memory, understanding, deductive logic, and induction). The emotions (as already noted) are a fundamental part of the process that determines the final decision. A mood or a memory that produces anxiety, happiness, or sadness could determine the type of decision that is made when facing a problem. Ultimately, decision making is reduced to the solution of a problem—searching for possible responses and choosing one of them. As is discussed in chapter 4, chess provides an ideal laboratory for the study of decision making. Move after move, the chess game is a sequence of decision-making events, and the player's task is to limit the responses—to reduce the set of legal moves down to a smaller set of good moves to find the optimum move. In this respect, the attitude that has been attributed either to Richard Reti or José Raúl Capablanca (two of the best classic chess players) is enlightening. When the player was questioned about how many moves he saw in a position, his answer was blunt: "One, the best one."

All possible solutions to a problem are, in theory, in a voluminous *search space*. The mind elaborates strategies for diving into that search space, organizing the best solutions for a certain problem in a way that allows the subspace of more desirable solutions to be considered with more zeal. In essence, that is the base of expert knowledge and the foundation on which programs to play chess are constructed.

In daily life, any situation offers us the opportunity to use all our cognitive processes to find solutions to the most common problems. For example, a woman is rushing to get on a certain train, but first she needs to cross the street with her child in a baby carriage. No cars are passing, but the traffic light has just turned red. If she waits for the light to change, she will not arrive at the station in time and will miss the train. If she crosses when the light is red, it could be dangerous, but she will get to the station on time. Her possible solution space includes four options—cross with her child, cross without her child, wait for the light to change, and return home. Option two (cross without her child) is ridiculous, so she does not take it into account. Option three (wait for the light to change) does not offer any benefits, since if she waits she misses the train and since there is no other train until the next day, it would be better to choose option four (return home). So the solution space is reduced from four theoretically possible options to the two best options—to cross or go home. To reach this *reduced solution space*, she mentally elaborated the following: (1) there is still time to catch the train if she crosses on the red (inference: the traffic light always takes ten minutes to turn green, and there are only five minutes left to catch the train; deduction: a simple calculation convinces her that there is no time if she waits); (2) if she does not cross on the red, she will miss the train (logical deduction), and there are no trains until tomorrow (reference from long-term memory). Now she must make a decision: cross or return home? A multitude of processes enter into play—the sense of danger and the type of emotional response that it supposes, her perception of the street and assurance that there really is no car coming, the risk she runs with respect to the true utility of taking the train, and so on.

Decision making is a rich cognitive facet and is one of the activities where the human species excels among the primates. In addition, throughout the sequence of her decision making, our woman has had to resort to something special—reasoning through thought. She generated images to reduce the solution space and to make a decision by using language, which we could call the *operative vehicle* of thought. One of the great evolutionary conquests of our species is the appearance of language as a novel characteristic, which has allowed us to generate levels of abstraction much greater than any other species. Moreover, language is so absolutely necessary to elaborate a thought that, without it, we could only generate images in our mind as direct representations of the world, without the possibility of abstracting them to carry out more elaborate cognitive processes.

Language

Cognitive capacities (like memory, decision making, and perception) are shared to a greater or lesser extent by numerous mammals, especially the primates. Nevertheless, symbolic language distinguishes us from other living beings and was acquired at some time in our evolutionary past as a result of a series of events that involved the body's anatomy and the brain's structure. Symbolic language has allowed human beings a development that is unprecedented in natural history in terms of our capacity to modify the environment.

Language is one of the clearest manifestations of the singular cognitive capacity of the human being. It provides a medium for carrying out acts of communication efficiently. The semantic transmission that takes place when expressing an idea transcends the mere representation of the world and enters into the sphere of desires, feelings, and moods. Its wealth of meaning and the ambiguities that those meanings entail confer on language unusually vast properties as a vehicle of the mind. The organized structure of language syntax has led some to postulate that the human brain is innately possessed of a suitable structure for learning a language.

Emotions

To reiterate, all human activity is influenced by some type of emotion (a point also highlighted repeatedly in Jonathan Rowson's *The Seven Deadly*

Chess Sins, where he emphasizes that emotion is the key to decision making). The perception of physical reality (through the feeling of skin, muscles, or any other structure of our body) also has an effect on the emotions themselves. Patients with cerebromedullospinal disconnection (locked-in syndrome) cannot move or feel their bodies (their only means of communication with the outer world is through vertical movements of the eyes) but have full consciousness of what is going on around them. Under such circumstances, it would seem that the patient would feel emotionally despondent. Nevertheless, apparently it is not so. The loss of the perception of oneself prevents the manifestation of emotions. Thus, both physical movement and the sensation of emotion play determinant roles when an intellectual activity is taking place. Think about the student who bites her nails during an exam (to think better?) or a lover desolated by a disappointment in love who curls up in bed in a fetal position (to protect himself from the world?).

The woman who had to cross the street with a baby carriage faced a dilemma where an important role was played by emotions—the fear of being run over, the desire to protect the baby, the anguish of not being able to arrive on time, the remorse of not having left home earlier. In chess, there is a whole repertoire of gestures and movements that accompanies the thinker. Players move their fingers simulating the movement of a piece, scratch their noses, rub their foreheads, bite their knuckles, jiggle their legs, or pace while thinking out the move. And emotions play a determining role in that a player who is emotionally engaged with a game will understand it far better than one who is not. If the mood is not propitious, the capacity to understand a position and to visualize a decision tree will be reduced considerably. But what role do emotions play in decision making? Some researchers (such as neurologist Antonio Damasio) look to emotions as the key to thought: without an emotional component, a decision cannot be made. If there is no objective behavior, if objectivity has an emotional component, or if emotions are elements that are not essentially different from other cognitive functions, then emotions can be studied as if they were ordinary brain processes. Can the chess player come up with his plan without feeling anything? The simple answer (with a deep meaning) is no. This simple answer is more than enough to distinguish a computer (whose mission is to calculate millions of positions per second) from a chess player (who is filled with emotions, sentimental memories, hopes, pain, happiness, and an endless variety of feelings about himself and the world he lives in). The player is conscious of the catastrophes that ravage the world (war, famine, inequalities), but machines are fully indifferent to them.

Consciousness and Qualia

Homo sapiens sapiens, the scientific name of the current human populations, is popularly claimed to be the only animal that is conscious of being conscious. This apparently unimportant idea carries deep connotations that throughout civilizations have produced art, mystical and religious fears, and the search for scientific knowledge. To be conscious of consciousness supposes taking an introspective position as an observer of the world that surrounds us. The consciousness of being conscious highlights what is social within us. Being aware of past, present, and future in an environment where other individuals exist grants us the capacity to know ourselves as unique individuals who are different from those around us.

The fundamental problem of consciousness resides in finding out how the brain provides the bases that are necessary for representing itself to itself. The experience (personal and interior) of being emotionally moved by a given situation (or simply reacting to any daily event—such as hearing the radio at seven in the morning, feeling the flow of water on our body when showering, or listening to a bird fluttering through the branches of a eucalyptus tree at dusk) generates very particular mental states that are called *qualia* (like *quality*). Qualia are witnesses to the consciousness of being conscious and, as such, are the philosopher's stone of theories about the mind. An epistemological problem with qualia is whether they can be studied scientifically or not. Since they are personal and within ourselves, they cannot be directly observed. (This is not a problem for science: think about particle physics, which studies the components of matter by observing disturbances on given experimental conditions.) In addition, one subjective experience does not necessarily have to be and in fact, is not, the same as or similar to another one. Thus, when observing the same scene, thinking about the same concept, or smelling the same rose, two different people experience different sensations (qualias) that are hard to compare. Another more serious problem is the ontology of qualia themselves. Do they truly exist? Or is this just another philosophical concept that does not add anything to the study of consciousness? We will leave this question floating out there for anyone who is interested in diving in to the fascinating and slippery world of philosophy of the mind. In any case, qualia point to the existence of intrinsic characteristics of the mind that need an explanation that, in the best possible scenario, would be based on the way that the brain functions.

The concept of consciousness includes various phenomena, which means that consciousness cannot be talked about in a scientifically valid way as if it were a global property. Consciousness is a modular phenomenon that includes different mental states and states of consciousness. According to Gerhard Roth, there are connections between the frontal cortex, the basal ganglia, and the thalamus that produce conscious actions. Emotions (the amygdala) and memory (the hippocampus) are the modules that provoke voluntary actions. Past experience and the evaluation of the situation are the indispensable requirements to carry them out. Where previous experience exists, consciousness is not necessary to carry out an action. It is automated and takes place in a way that allows the brain to be occupied with other activities. Only when the brain is confronted with a new situation does the conscious act appear. Thus, based on his previous experiences, a chess grand master has at his fingertips opening moves, all their variations, and thousands of patterns that he recognizes on the board, and therefore he is able to make a move automatically (see chapter 4). Finally, memories of things known from the areas of the neocortex are transferred to the internal subcortical areas, allowing actions to be executed automatically. A person accumulates facts and generates categories and generalizations that are increasingly global and that are housed in the subcortical memory areas, automating behavior with respect to everything that is related to these categories. Perhaps it is a question of generating more and more abstract hierarchical modules that are filled with facts (or that are simply prepared to be filled with facts). The more we know, the more capacity we acquire to relate knowledge, and the easier it is to solve a specific problem, since we have more resources to reach the looked-for solution.

To close this section, there are two concepts to look at that, according to Gerald Edelman, could be behind the generation of the consciousness of being conscious—*perceptual* and *semantic bootstrapping*. Edelman proposes two types of consciousness—*primary* and *higher-order consciousness*. Primary consciousness allows us to know the facts that we perceive, while higher-order consciousness puts the facts in personal perspective—within the idea of oneself, with a certain past and history, and with the consciousness of being conscious. The idea rests on the capacity of the brain to generate representations of the world by means of the formation of *neural maps* that consist of groups of neurons that are connected by synapses. The repeated action of a perceptual stimulus (*autonomic potenciation*) generates these maps, which end up constituting organized memory banks. Perceptual categorization, memory, learning, and a system of values that relates the stimuli of the world positively or negatively to the basic systems of survival (such as hunger, thirst, and sexuality) constitute the basic elements for the acquisition of higher-order cognitive capacities and conscious experience. To reach this, semantic self-generation and self-potentiation (the repeated action of a meaning for a specific fact) are necessary. Language plays a determinant role here, starting with the assigning of phonetic sounds to concrete meanings. In this way, brain maps are able to represent the world semantically and pass, for example, from familiarity with an image to its meaning. In Edelman's schema, the consciousness of being conscious emerges through semantic bootstrapping, which cannot be done without perceptual categorization, memory, and learning. This semantic organization of the world generates concepts—among them, the concepts of being and of the past and future in relation to the primary consciousness.

Intelligence

It is not easy either to speak about intelligence as a recognizable and measurable characteristic of animal behavior. Instead, it is easier to speak of intelligent behavior—that is, a type of conduct that requires the action of cognitive processes (described in the previous sections) and of many other factors. However, a behavior is directed toward some type of objective or forms part of the reaction to a given stimulus. So it is legitimate to ask which objectives require the use of intelligent behavior or what type of stimulus triggers the use of intelligent conduct. The answer to this question invariably rests on the appearance of novel events that the individual has not encountered previously. To be able to solve a problem related to a new stimulus, all the mental processing modules need to be activated: perception is necessary to understand the event. Memory enables us to see if we have ever had a similar experience and to identify the current one as new or not. Attention allows us to focus on the problem and to have the rest of the cognitive resources concentrate on solving it. Evaluation provides a feedback between recognition of the problem and the type of feeling that it provokes in us (such as, is it worth the trouble to solve it?). In addition, the motor action module prepares possible solutions to the problem within our mind and keeps them ready to order an appropriate action.

But an individual's intelligence undoubtedly undergoes changes throughout life. Thus, the type of intelligence that we have just described is known as *fluid intelligence* to differentiate it from *crystallized intelligence*. Fluid intelligence is developed essentially during the first years of life and refers to the ability to reason in relation to the solution of new problems. Fluid intelligence uses abstraction methods where experience cannot help with problem solving. Crystallized intelligence is consolidated during adolescence and young adulthood and refers to the capacity to reason founded on the wealth of experience accumulated in a given cultural environment. After twenty-five or thirty years old, both types of intellectual capacities begin to decline, although fluid intelligence falls with much more rapidity than the crystallized type. To some degree, the theories of intellectual development echo the Spanish proverb, *"Más sabe el diablo por viejo que por diablo"* (The devil knows more from being old than from being the devil)—roughly, "With age comes wisdom."

Three schools have tried to generate theories about the nature of intelligence. The psychometric school is based on measures of intelligence (that is, tests that measure the *intelligence quotient* or IQ), the school of cognitive psychology tries to explain which mental processes underlie intelligence, and the biological school tries to correlate brain function with intelligent activity. All the psychometric theories have been based on the idea that it is possible to measure the phenomenon of intelligence in some way. One theory about the nature of intelligence identifies a global organizing center g for the specific cognitive functions (Charles Spearman's theory). Another theory identifies groups of functions that are less inclusive than the global organizer (primary mental abilities such as visualization, spatial rapidity, verbal understanding, fluidity, numerical abilities, inductive reasoning, memory, and creativity) (Louis Thurstone's and Joy Guilford's theories). And a third, hierarchic theory includes the other two, with the organizing center g always intervening to generate an intelligent behavior that controls intermediate factors like verbal and nonverbal ability, which then in turn control specific abilities (figure 2.6).

Cognitive psychology has followed a different direction, considering intelligence as a set of mental representations and a series of processes that operate on these representations that allows the individual to adapt to the changing conditions of the environment. This type of approach is connected with information theory. The intelligent mind operates by processing information that it collects from the environment, and the better and faster this information is processed, the more intelligence is demonstrated. As was noted in the preface, artificial intelligence was born out of the relation between psychology, information theory, and computer science. In fact, one of the pioneering researchers in cognitive psychology is Herbert A. Simon, a Nobel Prize winner in economics, who was an active researcher along with Edward Feigenbaum, John McCarthy, Marvin Minsky, Allen Newell, and others during the beginnings of the science of artificial intelligence. Simon carried out the first research and theoretical models on expert behavior in chess and general models for problem solving (on which more later). Cognitive psychology has been nourished by ideas coming from artificial intelligence (and vice versa), such as the connectionist models and frame theory of Minsky and

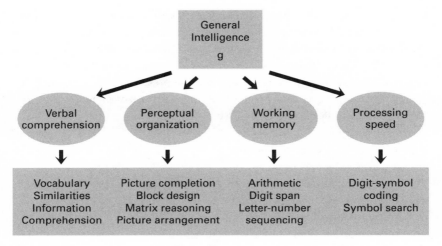

Figure 2.6

Hierarchy of the differences in mental abilities among individuals based on the Wechsler (Wais-III) test, an intelligence test for adults. Both working memory and processing speed (two fluid factors) are good candidates to explain the nature of g.

the parallel processing idea from David Rumelhart and James McClelland (which postulates that cognitive processes are carried out simultaneously at the moment of solving a problem and not serially, one after another, as classic theory postulates).

A fundamental problem for any theory of intelligence is to differentiate between individual differences and social influences. The same intelligence test can give very different results in different social contexts. To palliate these problems, cognitive theories have arisen that account for the context in which an individual lives to evaluate his intelligence. One of the most radical contextual theories was put forward by Howard Gardner in 1983. His theory of multiple intelligences postulates that, at a minimum, the following seven types of intelligence exist—linguistic, logical-mathematical, spatial, musical, bodily-kinesthetic, interpersonal, and intrapersonal.

The global space of operations model (described above) postulates the existence of a circuit of integration that puts processors or general modules of perception, memory, attention, action, and evaluation in contact. For cognitive tasks where the use of only one of these modules is not enough, the global circuit becomes activated and puts into operation parts of each module in a unique combination for each specific task. If we try to use this model to provide a brain and mental substrate that explains the nature of intelligence, we see that the global operation space that connects the modules can be identified as Spearman's factor g but not as a generator of commands. It is rather the other way around—serving when each of the modules turns it on. The modules can, in turn, be interpreted as the processes that are identified in cognitive psychology, where the presentation of the stimulus plays a protagonist's role as a mediator of the activation of each of these modules through the global space circuitry.

Born or Made? Genetics or Learning

There have long been debates about the percentage of capacity and disposition (to carry out a task, to reason, to memorize) that is innate and the percentage of these capacities that is gradually acquired thanks to a learning process. These debates are particularly important in the exploration of the differences between artificial and natural intelligence. If intelligence is a capacity that is gradually acquired as a result of development and learning, then a machine that can learn from experience would have, at least in theory, the capacity to carry out intelligent behavior.

The previous section showed that one school of cognitive psychology postulates the need to take into account the context (the social situation, the type of culture, the family situation) in which an individual grows and develops to avoid bias when it comes to determining his intelligence. Until the middle of the twentieth century, the way to determine the intelligence of an individual was through special tests that measure certain mental capacities. The father of this type of practice was French psychologist Alfred Binet, who is also the author of the first analysis of perception in chess, *Psychologie des grands calculateurs et jouers d'échecs* (*Psychology of the Great Calculators and Chess Players*) (1894). The important point for the moment is that Binet invented what today is well known as intelligence quotient (IQ), which he devised to analyze rigorously and quantitatively the intellectual age of a child in relation to his biological age. Binet specified that this type of test was to be used to help distinguish children with problems from normal children so that all children could be helped and given the means necessary for learning. It was not to be used to create a scale for normal children or to measure intelligence. In spite of these warnings from Binet, IQ has been used in an indiscriminate way in a multitude of situations that have ended up absurdly stigmatizing collectives. The problems with IQ and intelligence tests reside in the bias of their results, which are influenced by the social environment in which a child develops, which fosters or inhibits his intellectual development.

Learning is an activity that unfolds in all its fullness during the first stages of animal development. In humans, the first years determine the acquisition of faculties (such as attention, perception, and language) that, later on, will act as implicit parts in the mental machinery. To understand how one learns is to understand how those capacities are acquired during infantile development, and the Swiss psychologist Jean Piaget carried out a monumental work dedicated to understanding the development of the child.

Piaget postulated the existence of four phases of development—the *sensorimotor stage*, the *preoperational stage*, the *concrete operational stage*, and the *formal operational stage*. Essentially, a child begins life as a kind of empirical scientist who faces new situations and tries different strategies to solve and to understand them by trial and error (this is the sensory-motor stage, from birth to approximately two years old). In this way, children acquire the capacity to understand that there is a cause and effect relationship between one fact and another, and they begin to predict what will happen when carrying out some type of action. Later on, during the symbolic preoperational stage (between two and seven years), children are able to internalize the knowledge of the world and to codify it by means of structured symbols (language), which allows them to carry out mental experiments. Nevertheless, the capacity to generalize and to categorize groups of similar elements is not developed until the following phase, the concrete operational (from seven to eleven years old). In this phase, children are able to order and to group knowledge of the world into general categories and to understand cause-effect relationships in a more general way. Finally, in the formal operational stage, symbolism and the representation of the world become fully generalized, and predictions about the future take place on a scale that goes beyond knowledge of the present.

When passing through these stages, each individual confronts knowledge of the world in a personal and unique way. No two vital experiences are exactly the same, just as two people who are genetically the same do not exist. Even twins, who possess the same genome, show a high percentage of difference in its expression pattern (that is, the proteins that are expressed), since that depends on each one's unique interaction with the environment. As a result, learning determines, to a great extent, the possibilities of acquiring a more or less profound understanding of the world, depending on the contingencies of life. Intelligence is acquired gradually, and the family (and the scholastic environment) determines the development of cognitive capacities, which lay the foundation for the possibility of learning. No genius learns from ignorance. Nevertheless, certain personal characteristics in our genetic inheritance predispose us to a better or worse capacity to acquire and develop cognitive capacities. The existence of precocious children in certain disciplines (among them chess) is evidence that genetics also plays a role in the development of human intelligence.

Summary

The cognitive processes contribute to how we are who we are, how we believe that we are who we are, and how we know that others also know about our knowledge. The integration of the brain, the mind, and the body generates a person—an individual who grows within a cultural environment that shapes him at every moment. Decision making is the key element of our relation with the world, and we cannot escape it or ignore it. Day after day, hour after hour, minute after minute, the neurons are connected and disconnected. They transmit impulses, distribute the transmission throughout the brain and body, and create a set of cognitive processes that, integrated, constitute the mind. Our capacity to create, invent, suggest, evoke, carry through, love, feel, enjoy, laugh, cry, be astonished—really, our capacity to exist as people within a society—depends on the way that we modulate the mind (the states of consciousness) in a coherent whole that we store in the form of memories. We are intelligent beings insofar as we can access ways of making decisions to solve the daily problems that life presents us with, and that intelligence unfolds in infinite qualities, from logic to emotion. Each person processes information from the environment and reconfigures it using different doses of each one of these singular abilities of *Homo sapiens*. The mind creates a metaphor of ourselves and of the world that surrounds us. And it is so skillful that it has created machines that are capable of simulating human beings' own creativity in a series of 1s and 0s—the challenge of the following chapter.

3 Artificial Intelligence: Silicon Metaphors

In 1950, five years before the first basic developments in the field of artificial intelligence, Alan M. Turing, the English mathematical genius who was persecuted by his country's establishment for revealing his homosexuality, proposed the following scenario. There are three people: A is a man, B is a woman, and C is a person whose sex is immaterial (the scenario proposed by Turing would provoke a grin of understanding if his faithfulness to his own nature did not have such a tragic end). C can see neither A nor B, and by means of written questions (a teletype was used), C must distinguish the man from the woman by the differences in their answers. Their answers may be true or false.

In Turing's second scenario, the man A is replaced by a computer, and the analyst C must distinguish the machine from the human being. This test (now called the Turing test*) allows a more ambitious question to be answered: can machines show intelligent behavior? And that question reveals the hidden objectives of artificial intelligence—the recognition of the existence of something singular in our species that we do not see granted to other living beings and that we are driven to reproduce.*

The Silicon Metaphor: Computer Information Processing

A multitude of new concepts, new words, and household computers have been created as a result of the development of computer science. Actions such as scanning and printing a document, downloading a file from the Net, and defragmenting the hard disk are everyday currency, even in my parents' house. Behind all this is a history of discoveries and inventions that has brought the computer into the home but that originally pursued very different objectives that are directly related to the subject of this book. This chapter offers a brief history of computer science from its beginnings to the present time. Its approach is from the point of view of information processing as a basic foundation for artificial intelligence systems.

Almost immediately after computers appeared in the 1950s, they were compared to the human brain. This analogy was used in the scientific world (which developed disciplines like cybernetics and artificial intelligence) and in literature and film (which found these new information-processing machines to be sources of inspiration). The comparison between computers and the brain can be made without much difficulty. The fact that these machines carry out activities that traditionally have been considered to be cognitive functions indicates that a certain level of analogy exists between the two (or that those activities should not be considered cognitive functions). The classic example is calculation. When doing a moderately complex calculation (such as dividing one multidigit number by another), we use cognitive resources such as long-term memory ("how do you do division? how much is three times eight?"), working memory and attention ("twenty-four minus fifteen, cross out the two, carry one"), logic ("if I multiply by three the resulting number is too great, so it must be by two"), and so on. When a person carries out an activity of this type—making a rational and organized use of cognitive resources and of the facts and experiences accumulated in memory—we call it *intelligent behavior*.

What happens in a machine when it carries out this same activity? If a brain generates a certain mental state to do a calculation, can it be likened to the state that the machine is in when it does the same calculation? When a computer plays a game of chess, is the machine in a state that is equivalent to the mental state necessary for a person to play chess?

When the problem is framed this way, it becomes a question of processes. What matters is not the means of physical execution of the problem, but rather the mode (the type of state)—which might be equivalent even though it materializes by different means. An example from theoretical biology can help to generalize the idea beyond the field of artificial intelligence. In Chicago in the 1930s, Russian biophysicist Nicolas Rashevsky suggested that the *principle of biotopological equivalence* was a unifying law for all living things and emphasized the equivalence of processes and functions among different species. The process of digestion, for example, is essentially the same for all consumer organisms, from one-celled animals to vertebrates. The equivalence is one to many (to use the mathematical term)—that is, the digestive system of a paramecium is significantly simpler than that of a mammal, so that for each equivalent function there are a smaller number of structures that carry it out in the paramecium than in the mammal (figure 3.1).

To look at the idea from another angle and consider artificial systems as candidates for the equivalence principle, we need to analyze computers' structure. If the computer were organized in the same way as the brain and

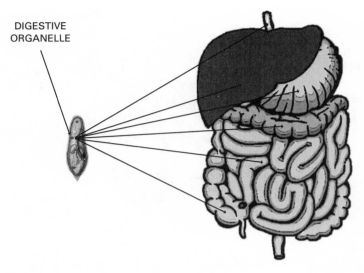

DIGESTIVE
ORGANELLE

Figure 3.1
Digestion and Rashevsky's principle of biotopological equivalence. *Left:* Diagram of *Paramecium caudatum. Right:* The human digestive system. Despite carrying out the same function, the human system is structurally much more complex. Arrows indicate this one-to-many relationship between the digestive organelle of *P. caudatum* and the digestive apparatus of a human.

in addition carried out the same activities, we would have to conclude that both entities must enter into states that are equivalent when they do those activities. That's an extreme conclusion that some artificial intelligence researchers would be prepared to sign off on. For example, Thomas Ray, a pioneer in artificial life research, is convinced that his programs are as alive as any organism. If human-built machines were included, therefore, the principle of biotopological equivalence could be extended, losing its biological exclusivity (figure 3.2).

But how far does the resemblance between the organization of a computer and a brain go? In chapter 2, the brain is shown to be a conglomeration of neurons that form communications networks organized in modular and functional areas. There is a logic to the brain's functioning as a center for storage and processing of information from the exterior (and interior) of the organism. This information comes primarily from the sensory organs, which convert the stimulus (such as sound waves) into electrical impulses. As a result, the brain responds to the stimulus, sending an electrical signal through the motor neurons, which generates some kind of action. This is

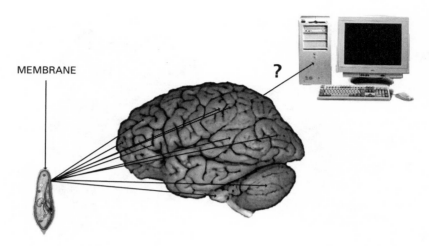

MEMBRANE

?

Figure 3.2
The possible extended biotopological equivalence principle. The membrane of the paramecium is a simple structure, relating the animal's inside with the environment. The complexity of a brain or a computer also relates inside and outside.

generally mediated by the *interneurons*, which act like communication wires between the brain and the sensory and motor neurons. In this way there is a well defined path of action—stimulus, sensory organs, interneuron, brain, processing, interneuron, motor neuron, response. We also saw that the stimulus can originate within the brain (as when we think) or can come from the outside. The same is true for the response. It can be a motor response or a thought, in which case the action path after processing would stay within the brain, generating something that could be considered an *interior action*. Let us see how far we can carry the analogy to computer architecture.

In a typical personal computer, there are components to input, store, process and output data. Input elements include the keyboard, mouse, floppy drives, memory cards, portable hard drives, microphones, touch screens, virtual reality helmets and gloves, and a long and growing list of others. The data-storage systems are called *memory*—either read-only memory or memory that can be recorded over again and again (such as on floppy disks or hard drives). Data processing takes place in the *central processing unit* (CPU), which is formed by a series of modules composed of *silicon circuits* (*chips*). Finally, output elements include screens, speakers, printers, and other peripherals. The passage of electrical current allows data to be transmitted between circuits, so all these components are connected by cables (or buses) that allow for communication among them.

The analogy between the brain and the computer holds at this superficial level—electrical current as a means of communication, wiring, networks or integrated circuits, transduction of external impulses (for example, the mechanical movement of the mouse is converted into electrical current), functional modules, data that is stored in memory, and output signals that respond to an input stimulus.

Nevertheless, a closer comparison between brain and machine results in a less promising metaphor. For example, how circuits are organized deserves attention. Each chip is made up of miniaturized silicon cables that allow the passage of current and that are organized in a way that allows one *logic gate* after another. Each logic gate is designed so that, depending on the type of information that the gate receives, it opens or not, preventing or allowing the current to pass according to the type of door. There are four basic types of logic gates—NOT (inverter), AND, OR, and XOR (exclusive OR). The operation of these gates corresponds with the elements of Boolean logic. Thus, in NOT (inverter) gates, whatever the type of information, the output will be the inverse. In AND gates, if both inputs receive current, then the output will allow the current to pass. In OR gates, if only one receives current, the output will also allow the current to pass. And finally in XOR (exclusive OR) gates, the current will be allowed to pass only if one (not both) of the inputs is on; if both inputs are receiving current or neither one is, then there is no current output.

If logic gates are compared with neurons, it is clear that the latter are much more complex entities since the decision to allow the current to pass or not is made after a threshold stimulus has been reached, which comes after evaluating the sum of stimuli from a large number of synapses (possibly some 10,000). In addition, the current's passage (the action potential) is modulated chemically, so that it also depends on the type of neurotransmitter that is released in the presynaptic neuron.

Another fundamental difference resides in the type of information that is processed. Computers process binary digital information in such a way that all input data, output data, and data that is stored in memory are codified as a series of 1s and 0s, packed in larger information units called *bytes* (sets of eight binary digits). This means of manipulating data is especially apt for use on electrical circuits: 1 represents the current flowing, while 0 represents its absence. Thus, each byte is sent as a sequence of eight pulses where the electric current is on (1) or off (0). Data storage in computer memory is based on packets of bytes that represent information such as a number, word, or image. The information is accessed thanks to the directions that the system has for each packet of bytes, so that each record can be quickly found just by recovering the address where it is stored.

Within the brain, things are very different. The first great difference is codification. As has already been shown, memories are stored at the cellular level in the form of a framework of synapses and at the molecular level in the form of proteins that possibly are distributed throughout a large number of neurons. Another fundamental difference between computer systems and the brain is the recovery and fidelity of stored memories. Computers have effective systems of memory recovery that call up the proper location and pull up exactly the same record that was stored. The brain possesses a series of synaptic maps that cannot always be recovered exactly. Human memory is variable and, in many cases, incomplete and even erroneous. The brain has neither 0s, bytes, nor precise addresses for each memory but rather a series of synaptic networks that strengthen throughout time as a result of a series of biochemical and cellular mechanisms (such as long-term potentiation, discussed in chapter 1). Even worse, our memory is altered by our experiences to such an extent that on numerous occasions what we remember is something that we have reconstructed based on what we know, what we really remember, what we think we remember but we invented, and an endless variety of other variables that are precisely what makes us human, marking our difference from computers.

But the most noteworthy difference between a brain and a computer is *functional plasticity*. The brain has the capacity to reconstruct its synaptic connections and to adapt them to very different functions. The fact that every day we learn and remember something new demonstrates that. From birth on, the brain's memory maps are continually taking shape, creating representations of reality that are so faithful that at times they seem as real as the present that we are experiencing. Moreover, the brain's plasticity is so great that it can be reorganized and generate new specialized centers. An extreme example is that of epileptic patients who have had a large part of one of their cerebral hemispheres extirpated. In numerous cases, the patients reorganize the remaining hemisphere to take on the functions that initially were carried out by the joint work of both hemispheres. The Broca and Wernicke areas in these patients no longer exist, yet they are nevertheless able to speak, read, and write using only the right hemisphere. Thus, not even the phenomenon of lateralization (chapter 1) is determinant of brain organization and functioning. Rather, plasticity seems to be the key to the success of vertebrates' brain architecture, which is the opposite of how computers operate. A computer's structure literally demonstrates the adjective *hardwired*:

no matter how flexible its software is and how much a computer can learn, it will still be subject to the rigidity of its architecture.

Perhaps a day will arrive when we can construct a computer whose wiring can change in accordance with its experiences. In fact, experiments in which neurons have been made to grow within silicon circuits to form a hybrid connection between a computer and an organism have already taken place. But until then, this fundamental difference between the brain's neural networks and a conglomeration of chips seems an insurmountable obstacle to building a machine that can carry out intelligent activities. The type of intelligence that can be simulated depends on the software and not the hardware, with the consequent limitations that supposes for emulating human behavior. The field of artificial intelligence has followed this path, creating programs that use programming tricks to simulate neural plasticity, evolving toward robotics approaches in which the embodiment of the software in appropriate hardware is crucial.

A Protohistory of Artificial Intelligence: A Web of Desires and Ideas

Artificial intelligence as a scientific discipline dates its more or less official origins to the middle of the twentieth century, but various earlier ideas and devices relate to AI's earliest stirrings. Still, its development is closely tied to that of computer science. Certain notions, such as the Golem or Victor Frankenstein's creation, indicate a common human desire to control life and generate its properties. Through the pioneering contributions of Leonardo Torres y Quevedo, chess also shares a protaganism in this move to simulate life by mechanical or electronic means. There are at least four noteworthy myths that are part of this race to understand and dominate the life force— myths that are based on desire, necessity, curiosity, and power.

Myths of Desire
What can be said of desire? We already know that love can move mountains and generate life from inanimate matter, or at least so it is told in the classical myth of Pygmalion and Galatea. In book X of Ovid's *Metamorphoses*, Greek king Pygmalion renounces the love of women, carves an ivory statue of a woman, and falls in love with his own creation. The statue is given the gift of life thanks to the intervention of the goddess Venus (the Roman Aphrodite). George Bernard Shaw reworked this legend in his play *Pygmalion* (1913), and Alan Jay Lerner and Frederick Loewe's *My Fair Lady* (1956) brought Shaw's story to Broadway and the cinema. This story represents a departure point that has fed the imagination of cultures and societies using legends, just as

has the Biblical creation of Adam out of shapeless mud. Divine intervention is fundamental to creating life in both Judeo-Christian and Greek mythology, which is why it is not completely correct to include it as a predecessor of artificial intelligence. In the myth of the Golem, a human recreates life, although still as a mediator between the divine and the earthly. It is only starting with Mary Shelley's *Frankenstein*, which springs from the nineteenth century's fascination with science, that life is generated by physical means thanks to human talent rather than divine intervention.

Myths of Necessity

Necessity goes hand in hand with the legend of the Golem, which in the Ashkenazic Jewish tradition presents a sculpture that comes to life, this time not with loving aims but rather as a source of defense against the endless persecution of Jewish communities in Europe. The story tells of Loewenstein, a famous rabbi in sixteenth-century Prague, who, alarmed by the constant attacks from Central European Christians on his Hebrew neighbors, decides to create a creature from a mud sculpture, giving it life thanks to a series of cabalistic statements (divine intervention). The Golem is the protector of the Prague Jews until, out of control, it must be returned to its inert condition. This story points to the dangers of artificially generating what should be left to natural evolution, an idea that has been repeated time and again in the history of thought. For example, nowadays one can find views that oppose genetic engineering as a new "threat" to the "natural order." Or, as we will see in the following chapter, the same debate has arisen over the possibility that a chess program could overthrow a world champion. In both cases, the debate is badly framed, and the answers revolve around the capacity of human beings to transcend nature, a capacity that as we live our animal lives is a singular and indeed our best characteristic.

Myths of Curiosity

There is no more powerful force for human creativity than curiosity. Much of what we do springs from the insatiable curiosity that begins in our childhood, when we want to know, tell, say, try, ask. In chess, we are driven to find out if our proposal is a good one or if we should have tried a different move or another idea. In this protohistory of artificial intelligence, a second warning of the dangers that might lurk behind the conquest of the secrets of life comes from Mary Shelley's *Frankenstein; or, The Modern Prometheus* (1818), in which Victor Frankenstein's curiosity leads him to experiment with body transplants and electrical charges to generate life from a corpse. But an error is made (the creature is given a brain from a criminal instead of a scientist), and the story ends badly. Again, the message is that playing with what we don't understand can be dangerous.

Myths of Power

Another interpretation of the Pygmalion myth is Austrian director Fritz Lang's excellent film *Metropolis* (1926). In this movie, the beautiful protagonist is replaced by a perverse mechanical replica that generates chaos among workers who are exploited by mechanization and the assembly line. This film looks at how technology controls the means of production. Life is created to maintain power. On this occasion, there is no divine action in the machine's vital transformation. As with *Frankenstein*, the conquest of the secret of life is obtained thanks to human technical capability. Power is a force that appears throughout civilizations: wizards and shamans cure the members of their tribes and seers read stars and predicts eclipses. Astonishing and exalting, knowledge engenders dominion.

There is a common denominator in these four stories (Pygmalion, the Golem, *Frankenstein*, and *Metropolis*), and that is human beings' desire to control the biotic processes, generate life from inert matter, and transcend the restrictions imposed by time on their passage through the world (figure 3.3). Humans have long been obsessed with transcending death and securing eternal life. Oddly, in the Judeo-Christian tradition, curiosity for the fruits of the tree of knowledge condemns the human species to a finite existence. Are we rebelling against divine fate so that we can return to the natural state of eternal life—moving away from religious faith and approaching science as a means of liberation?

Science has in some ways crushed the gods and elevated itself as the means for transcending the physical bounds of life. But science is in fact a triumph of the human species and the creative capacity of our brain. If our ancestors had had our present-day knowledge and technological capacity, perhaps they would have written a science fiction novel instead of the Old Testament.

Science Fiction

Metropolis presents a story very much like that of the Czech dramatist Karel Capek's play *R.U.R. (Rossum's Universal Robots)* (1921). In this work, Capek coined the term *robot* (from the Czech word *robota*, forced work), which continues today to designate an automaton capable of generating some type of human activity. In both *Metropolis* and *R.U.R.*, the first industrial revolution's mechanization threatens workers and humanity as a whole. In the France

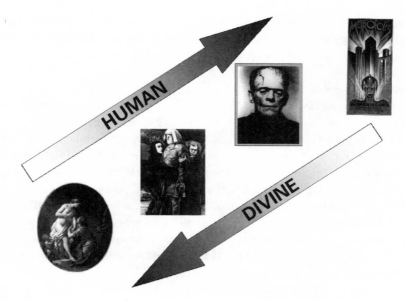

Figure 3.3
Four myths as predecessors of artificial intelligence—Pygmalion and Galatea (desire),
Golem (necessity), Frankenstein (curiosity), and *Metropolis* (power). The secret of life
is sought in a gradient of divine intervention to human intervention, which is also a
temporal gradient, representing the triumph of science over religion.

and Great Britain of the eighteenth and nineteenth centuries, workers
destroyed machines for fear that they would take away their jobs (England's
Luddite movement is an example).

A fluid interface between scientific and technological developments gave
rise to artificial intelligence and philosophical ideas about the nature of the
human species and its position in the world. These ideas have been fertile
ground for academic thought and for a multitude of science fiction stories
and films. Thus, in 1950 (the same year that Alan Turing created his test),
Isaac Asimov proposed three laws of robotics that have greatly influenced the
growth of the myth that AI is a possible threat for humanity:

- A robot may not injure a human being or, through inaction, allow a human
being to come to harm.
- A robot must obey orders that are given to it by human beings, except where
such orders would conflict with the first law.
- A robot must protect its own existence as long as such protection does not
conflict with the first or second law.

In the second half of the twentieth century, artificial intelligence as a conceptual field was nourished by scientific and philosophical ideas and by the imagination and ideas of writers in the science fiction genre (following nineteenth-century authors such as Mary Shelley, Jules Verne, and H. G. Wells) who went beyond actual scientific achievements and proposed ideas that would be considered in AI and robotics in a unique, successful collaboration between art and science. Artificial intelligence (and its sister branches robotics, cybernetics, and genetic engineering, which joins the first two like the guest from biology at this markedly mathematical and computational scientific party) resonates in one way or another with ideas about the possibility of controlling and replicating life itself.

A Protohistory of Artificial Intelligence: Scientific Elements

The history of AI is strongly integrated with other areas of science, particularly mathematics, neuroscience, and computational science, and modern AI advances in parallel with the spheres of knowledge of these other research areas. Aristotelian logic, for example, is the first essential link in this chain. Pythagoras's idea of the number as a generator of reality and Plato's idea of the model laid the foundation for formal metaphors about the world's structure and functioning.

With Euclid's *Elements*, the idea of algorithm is introduced, along with a deductive logical system for discovering geometric properties. In the thirteenth century, the neo-Platonic, Majorcan philosopher Raimond Llull, in his *Ars Magna*, used the algorithmic idea to generate a model to deduce truths (*ars inveniendi veritatis*) that are based on the combination of axioms.

In around 1642, the philosopher Blaise Pascal invented the first calculating machine (to help his father calculate taxes as the administrator of Rouen, France). Pascal's machine calculated by counting integers and is considered the first digital device. (The difference between digital and analog is based on the use and manipulation of discrete information, such as integers, rather than continuous information, such as real numbers.) Around 1673, German philosopher Gottfried Leibniz improved Pascal's machine by adding the capacity to multiply and divide to the addition and subtraction of the original machine. In the nineteenth century, English inventor Charles Babbage devised (but could never build) an *analytical engine*—a machine that could make analytical calculations and process digital information. As Martín Rasskin notes in his book *Música Virtual* (*Virtual Music*), "This machine was incredibly similar to current computers, except with regards to a slight difference of an ontological character, because it never existed."

Advances in logic systems—Boolean algebra (middle of the nineteenth century) and the symbolic formalism of Alfred Whitehead and Bertrand Russell (early twentieth century)—also contributed to the accumulation of elements that fed into artificial intelligence. The legacy of George Boole's algebra is particularly important, since it introduced a formalism that unites discrete logic with mathematics, an indispensable element for the development of computer science. All digital circuit construction is based on logical decisions that are formalized by means of Boolean algebra. The German engineer Konrad Zuse constructed a calculating machine in 1937 that was based directly on Boolean algebra, while Claude Shannon and John von Neumann recognized the importance of the binary system and Boolean algebra to construct digital information-processing machines. The relation between computers, cognitive processes, and artificial intelligence is made clear in a book by Boole whose title says it all: *An Investigation of the Laws of Thought.* The book, written in 1854, has been an important influence on the development of modern information science.

In neuroscience, Ramón y Cajal laid the foundation for the modern study of the brain and for AI when at the end of the nineteenth century he identified discrete units (neurons) that were the constituent elements of the brain's anatomy. (As is shown in chapter 1, until Cajal's work, it was believed that the brain was a mass of undifferentiated tissue.) Before Ramón y Cajal's work, researchers like Paul Broca and Carl Wernicke studied patients with brain injuries and identified areas of the brain that were implicated in cognitive functions such as speech. Finally, Donald Hebbs, who began working in the mid-twentieth century, proposed the first model of synapses between neurons capable of retaining information, thus completing the idea of the brain as an electrical circuit (see chapter 1).

The Modern Development of Artificial Intelligence

AI as a scientific concept was proposed by John McCarthy in a conference at Dartmouth College in 1956, and its development since then has been spectacular. In fifty years, AI researchers have experienced euphoria and pessimism. The euphoria springs from the undoubted attractiveness of AI's scientific objectives—to decipher the logic of cognitive processes and the mechanisms of the thinking mind. The pessimism comes when scientists encounter great difficulties when they try to reach these objectives. AI is currently enjoying an excellent moment thanks to the numerous ideas that it has generated related to understanding the brain's structure, the brain's functioning (a connectionist paradigm in cognitive science), and AI's practical applications in diverse industrial areas. Among these applications, the development of chess programs certainly holds a place of honor. Many of the ideas that have sprung from chess programming have been used to solve problems in very different areas, but chess programming, viewed as part of an epic "man versus machine" battle, holds a strong fascination on its own.

Other contributions much closer to the foundations of AI as a scientific discipline come from the work of John von Neumann's automatons theory, which provided a foundation for artificial systems of reproduction, and from Turing's ideas about a universal modeler that can make any kind of representation of the world. In 1943, Warren McCulloch and Walter Pitts proposed theoretical models of neuron operation as logical processing units (similar to the logic gates discussed previously). The work of McCulloch and Pitts is the base from which an entire field of AI using *neural networks* has been developed (figure 3.4). With game theory, Oskar Morgenstern and von Neumann proposed the use of the *minimax algorithm* to explore the *possibility tree* in games. This algorithm is used in diverse applications in AI, especially in chess (game theory and algorithms like minimax are discussed in the next chapter).

Meanwhile, Claude Shannon established information theory with his famous formula $H = -\Sigma p(i) \log p(i)$, where H denotes entropy, i denotes a specific event, and $p(i)$ denotes the probability that event i will occur. With this formula, the amount of information that a certain phenomenon has can be established. In fact, H measures the amount of uncertainty that exists in the phenomenon. If there were only one event, its probability would be equal to 1, and H would be equal to 0—that is, there is no uncertainty about what will happen in a phenomenon with a single event because we always know what is going to occur. The more events that a phenomenon possesses, the more uncertainty there is about the state of the phenomenon. In other words, the more entropy, the more information. Information theory was part of an enormous revolution in field research in telecommunications and therefore in the development of computer science and artificial intelligence. If we consider the brain as an amazingly complex information processor with an enormous capacity to generate different states (events), the metaphor of Shannon's formula seems to indicate a way to establish different levels of complexity in the development of the brain's cognitive capacities. As cognitive capacities are added during infantile development (this can be extended also to the addition of cognitive capacities during our species' evolutionary past), the amount of information and complexity of the brain increases, and more complex information from the outer world can be processed.

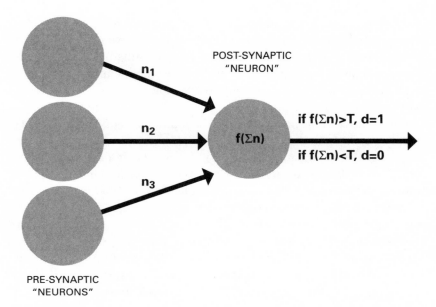

Figure 3.4
Theoretical model of a neuron, proposed by McCulloch and Pitts in 1943. This type of abstraction is the basis for building artificial neural networks. Presynaptic neurons fire an all-or-nothing stimulus (1 or 0). The postsynaptic neuron adds up the input values, using f as an evaluation function, making a decision about firing or not, according to a threshold value T.

A Protohistory of Automatic Chess

The creation of an automatic device that could take on a human chess player has captivated many generations. In 1770 in Vienna, Wolfgang von Kempelen, from Bratislava, constructed the celebrated Turk, a life-size, mustache-wearing automaton that he dressed in flowing robes and that was able to beat several famous chess players over many years (figure 3.5). Unfortunately, the secret of the Turk was under the table on which the chess board was placed. There, a Polish operator (named Boleslas Vorowski in the 1927 French film *The Chess Player*), who had lost his legs in one of the many Central European wars, moved a series of gears as required. Among the famous contenders who lost to the strong Polish master were Napoleon Bonaparte and Benjamin Franklin. The false automaton changed owners, cities, and chess players over several years until it was destroyed in the great fire of Philadelphia in 1854, not without first having been unmasked by the

Figure 3.5
A drawing of the eighteenth-century, chess-playing automaton called the Turk and a
photo taken by the author of a reconstruction in a café in the old-town area of Bratislava, Slovakia.

scrutiny of many observers (including Edgar Allan Poe, who published a
newspaper essay about its operation in 1836). There were other famous attempts to construct false automatons of this type, including Mephisto, which
was operated by electromechanical remote control by several top figures of
nineteenth-century chess.

The first real automaton that was programmed to recognize and carry out
chess movements without human intervention came from the Spanish inventor Leonardo Torres y Quevedo (figure 3.6). His machine, which can still
be visited at the Polytechnic University of Madrid, could start with its king
and rook in any position and reach checkmate on the opposing king. This
automaton is important in chess and in the development of calculating machines in general, including computers. The *heuristic* (the programmed rules
that generate a decision about an action that must be taken depending on
the state of the system) of the machine is based on a reduced series of criteria that must be evaluated whenever it is called on to move: it must (1) see
whether the rook is in danger and move it to defend it, (2) bring the king and
rook together to reduce the opposing king's mobility and see whether the
kings are in opposition, (3) check with rook, and advance a rank to reduce
mobility, and (4) repeat until checkmate.

Some Relevant Concepts: Recursiveness, Algorithms, and Heuristics

The programming of computer science applications has evolved at a vertiginous rate in its half century of existence. Operating systems and platforms

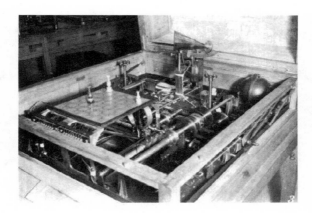

Figure 3.6
The chess machine of Leonardo Torres y Quevedo.

for the use of the computer itself, languages for programming applications for all tastes, and libraries with elements to help in programming are constantly changing resources. These elements respond to the exigencies of the hardware (which also advances at an amazing rate) and to the abilities of the industry and scientific research to see in computer science limitless possibilities to accelerate and deepen their respective knowledge domains.

Recursiveness and algorithms are two important concepts in this development. Both are programming tools that make clear the architecture of computers and the ways that they operate serially. Recursiveness as a programming structure is difficult to evaluate but possibly has contributed the most of any component to the rapid development of computer science. *Recursiveness* is the repetition of an event a given number of times until a condition is produced that ends the event. The events can be of diverse nature—carrying out a mathematical calculation, changing the color of screen pixels, generating sounds. Task automatization is possible thanks to recursiveness. Thus, a system that carries out a task (such as controlling a room's temperature) is basically a *loop* (the name for recursive structures) that has, for example, the following exit condition: "If the temperature is equal to or greater than 25 degrees, then turn on the air conditioner" (figure 3.7).

Programming is in essence a set of recursive structures—a conjunction of loops that interact by changing the values of variables and using parameters to control their starts and stops. Those loops, commands, and conditions are known as an algorithm.

```
WHILE (ON){
DO
{term=0);
verify_onstatus (term);
}WHILE (temp<25);
DO{term=1);
verify_onstatus (term);
}WHILE (temp>=25);
}
```

Figure 3.7
A recursive structure. The command DO is repeated until the condition WHILE is met. In this example (written in programming language C), the program gives the value 0 to the variable term as long as the temperature is less than 25 degrees, calling the function "verify on." As soon as the temperature increases above this value, the compiler jumps to the next recursive structure, giving the variable term a value of 1; this is interpreted as "turn on the air conditioning."

An *algorithm* (a word derived from the name of the nineteenth-century Arab mathematician al-Khawarizmi) refers to a successive and finite procedure by which it is possible to solve a certain problem. Algorithms are the operational base for most computer programs. They consist of a series of instructions that, thanks to programmers' prior knowledge about the essential characteristics of a problem that must be solved, allow a step-by-step path to the solution (figure 3.8).

An algorithm cannot reach a kind of information processing that resembles what human beings do. For that, strategy is needed that can use knowledge about a given situation as a base for decision making. *Heuristics* were created to do so. They are a means of programming that incorporates privileged information about the aspect that is being simulated as well as an operative manner of evaluating the information.

The fundamental characteristic of the heuristic strategy is a basic structure of programming denominated IF . . . THEN. These programming structures form decision chains called *productions* (although IF . . . THEN chains are also used in algorithms to make decisions at some point in the operation). For example, in the flow chart in figure 3.8, the chess program is using heuristics when it checks whether it is still within its area of knowledge about openings. In addition to productions, a program needs an evaluation module, which is in charge of making decisions (an *inference engine*).

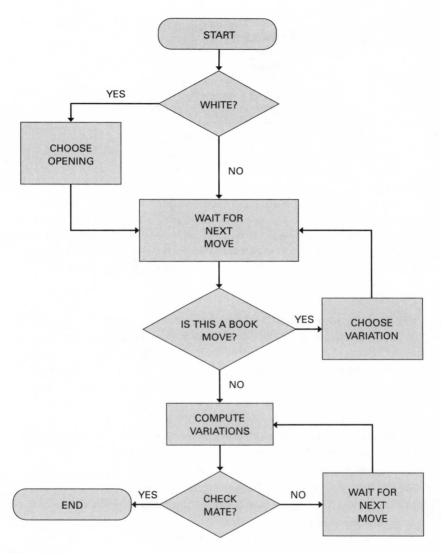

Figure 3.8
Flow diagram that solves a problem by following an algorithmic approach. This one shows simplified events that occur when a computer program plays chess. These kinds of diagrams simplify complex problems by making them into decision trees.

Thanks to privileged information and the different mechanisms for evaluating decisions, programming based on heuristics solves problems without having to cover the entire *possibility tree* or *search space* that an algorithm would have to do until it found the solution it was seeking. In chess programs, this is achieved by means of an *evaluation function* that is used for verifying how good each move is in relation to the board position. To a large extent, this function delimits the strength of a chess program. It is fueled by basic ideas of chess strategy and tactics. In addition to search algorithms for chess programs based on the *minimax* system, programs also use methods like *alpha-beta* that can efficiently search variations without having to waste time on those that are useless from a chess point of view (see the following chapter).

Expert Systems and Knowledge Engineering

Expert systems are a direct product of heuristic programming and have been developed to solve a multitude of complex problems, such as the diagnosis of bacterial diseases (MYCIN), the generation of molecular structures (DENDRAL), or any program that plays chess. An expert system is divided into a knowledge base, an inference engine, and a user interface. The knowledge base holds the facts that are known and the rules that connect facts among each other and is used to reach conclusions (for example, any IF . . . THEN production). In addition, the knowledge base incorporates new information that is contributed by the user to its permanently updated database. The inference engine chooses the rules of logic that are applied to the data and facts to arrive at the desired objective. The several strategies that are used to control this choice generate subproblems and solve them by applying the knowledge stored in the rules and contributed by the user.

In recent years, the term *knowledge engineering* has been used to refer to a part of artificial intelligence that particularly centers its objectives on the ways that human knowledge can be represented in a machine and on the diverse strategies that can be used to manipulate it. The recent successful applications of AI in industry, commerce, and education have been made possible by improvements in methods to codify and manipulate expert knowledge and by reducing the application of an expert system to a very concrete area where the knowledge and decision making can be limited and defined in an effective way.

Artificial Neural Networks

Artificial neural networks simulate at a basic level the structure and operation of the neural networks that are present in the brain. The parallel processing of brain function is simulated in a way that allows information to be distributed throughout all the network components. Essentially, the model connects several *McCulloch-Pitts cells* so that information can flow from one to another, moving from the input neurons to the output neurons (figure 3.9).

The simplest basic architecture of an artificial neural network is composed of three layers of neurons—input, output, and intermediary (historically called *perceptron*). When the input layer is stimulated, each node responds in a particular way by sending information to the intermediary level nodes, which in turn distribute it to the output layer nodes and thereby generate a response. The key to artificial neural networks is in the ways that the nodes are connected and how each node reacts to the stimuli coming from the nodes it is connected to. Just as with the architecture of the brain, the nodes allow information to pass only if a specific stimulus threshold is passed. This threshold is governed by a mathematical equation that can take different forms. The response depends on the sum of the stimuli coming from the input node connections and is "all or nothing."

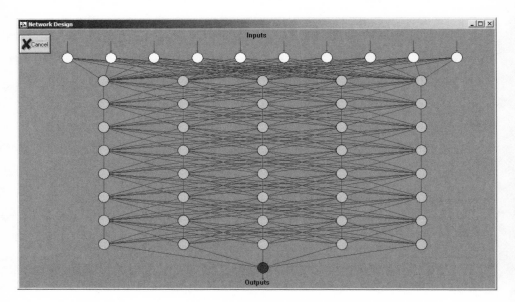

Figure 3.9
An artificial neural network with ten input nodes, eight intermediate layers of five nodes each, and one output node. The strength or weight of each connection can vary according to experience, extending the theoretical model of McCulloch and Pitts. These networks can simulate learning in ways that are closer to how animal brains learn. This network has been drawn using the commercial program Qnet.

The information in a simple perceptron or in a multilayer perceptron flows from the input cell to the output cell unidirectionally (*feedforward mechanism*). This architecture prevents the neural network from efficiently learning to recognize many patterns. As a result, artificial neural networks have evolved to foster the capacity of learning from experience, and new mechanisms have placed them closer to how the brain functions. For example, one of the mechanisms that these models use to carry out learning tasks is based on varying the strength of the connections between nodes as a function of the right answers through an algorithm known as *backpropagation*. The layers closest to the output make these strengths change, propagating the change toward the lower levels near the input layer. These dynamic changes make the network more capable of adequate learning. Only a few experiments have been made using artificial networks to create chess programs (see chapter 5). This scarcity is due mainly to the slowness with which a network can be trained to learn a problem as complex as the massive search for variations during the course of a game.

Elephants Don't Play Chess

Elephants don't play chess, but we do. Not only is there an apparently insurmountable distance between the complexity of bacteria and the complexity of Shakespeare, but there is also a big distance between two mammals such as an elephant and a human being or two cousin species such as any primate and us. What's special about humans that makes us show intelligence in a way that elephants cannot? More important for the present discussion (clues to the first question were advanced in the preceding chapters): what is so special about our intelligence that a bunch of symbol-processing algorithms cannot mimic?

As was discussed at the beginning of this chapter, the metaphor that compares computers and brains falls short of being accurate. Types of memory storage, communication dynamics, and learning plasticity are all very different for computers and for most classical AI systems. In addition, brains are a product of hundreds of millions of years of evolution, which sets them definitively apart from machines and algorithms. After realizing these and other related problems, several researchers embarked on an ambitious goal to generate bioinspired software and hardware to get closer to the nature of our fabulous brains.

Rodney Brooks, one of the founders of the modern approach to AI, wrote an article in 1990 entitled "Elephants Don't Play Chess": "There is an alternative route to Artificial Intelligence that diverges from the directions pursued under that banner for the last thirty some years. The traditional approach has emphasized the abstract manipulation of symbols, whose grounding in physical reality has rarely been achieved. We explore a research methodology which emphasizes ongoing physical interaction with the environment as the primary source of constraint on the design of intelligent systems" (p. 3).

Brooks was laying out the philosophical grounds of what has become known as, among other more or less fancy names, "nouvelle AI," "situated activity," or "strong AI." After the bold, original hopes of the founders of AI in the 1950s, both research and expectation about the real goods that AI was able to deliver began to cool down. In the 1980s and 1990s, however, new ideas started to emerge and refurbished some old AI concepts. In addition, increases in computer power afforded researchers with ways to tackle problems where complexity and combinatorial explosion were present. Simple search algorithms and heuristics became massive parallel search engines, expert systems gave rise to data-mining algorithms, and perceptrons and neural networks have been endowed with powerful techniques such as back-propagation—all in an effort to make machine learning a more efficient and close-to-human process. In the aftermath, this turmoil of effervescent new and exciting research has left two new concepts that have fueled the success of today's AI field—embodiment and agents.

Agents, Embodiment, and the New AI

From the beginning of AI, it was apparent that what was easy for humans (walking around without stumbling into a chair in our living rooms) was difficult for any artificial system. Too much information had to be processed in real time—seeing, recognizing, and reacting to the fact that a chair is in one's way. Conversely, what was difficult for humans (such as multiplying two twenty-digit numbers) was a piece of cake for a machine. This has been called *Moravec's paradox* (after Hans Moravec, the Austrian AI researcher who articulated this paradox in the 1980s). This conundrum sets the scenario for the appearance of bioinspired concepts that try to reduce the reach of the paradox, at least at the machine's end.

This new approach to AI includes genetic algorithms of several classes that make the software evolve by mimicking the processes that occur within genomes during evolution (such as mutations and crossovers) and the cooperative decentralized strategies for problem solving in swarms, cultures, and societies. These fruitful strategies have been used in robotics to model biological and intelligent behavior.

In the 1990s, a powerful new metaphor entered the world of AI. It changed the focus of software design from flat single-purpose algorithms to *agents*, a flexible computing entity that can be used in many different ways (thus highlighting the importance of modular design) and that can be engaged in massive parallel operations by cooperating as multiagents. From the collective behavior of these entities, complex new behavior occurs that is able to solve the original problem. Notions borrowed from complex dynamical systems are also part of the equation of this new paradigm. For example, the notion that agents can use a self-organized process to generate an emergent supraagent entity that is capable of solving new tasks is also an important issue in the new AI paradigm. In addition, the approach to programming agents and multiagents is a bottom-up approach that has no centralized control of operations but instead has a collective control that emerges from the individual behavior of each agent.

Finally, the notion of *embodiment* (related to robotics) reflects the basic idea that an intelligent system is trapped within the material and physical confines (including its form and size) of its structural body. This material dependence is not only relevant but determinant. This precisely encapsulates the postulates of this book, and it clashes head-on with the idea that it is the process, not the matter, that really gives rise to life processes, including intelligence. Rolf Pfeifer and Josh Bongard, in *How the Body Shapes the Way We Think: A New View of Intelligence*, provide an excellent overview of embodiment and its consequences for the future of AI. In their words, "Intelligence requires a body," a statement that prevents any piece of software from having or showing any kind of intelligence. Moreover, it puts back into the picture the idea that the evolutionary process (as briefly outlined in chapter 1) is an all-important force in nature, without which the mere notion of intelligence is meaningless (incidentally providing yet another argument to counter creationism; how many more are necessary to make it disappear from the public educational agenda?).

Summary

Many elements from very different spheres have come together in the story of artificial intelligence. The human needs to understand how we understand and to transcend our finiteness by conquering the secret of life are the forces behind the design of artificial minds. Four ideas are distant precedents of artificial intelligence and have been illustrated in the myth of Pygmalion and Galatea (desire), the legend of the Golem (necessity), the story of Frankenstein (curiosity), and the film *Metropolis* (power). Advances in logic, mathematics, and neuroscience nurtured AI until technology, through the development of computers, was ready to provide an appropriate platform for takeoff. Boolean algebra, Shannon's information theory, and the theoretical neuron model of McCulloch and Pitts laid a solid foundation for the rapid evolution of AI. There have been many advances since early expert systems and neural networks, and today notions such as agents and embodiment have AI cruising toward bioinspired systems in its continuous search for the nature of human intelligence. The third metaphor (created by the ingenuity of an extraordinary series of personages in the history of the arts, sciences, and thought) prepares us to receive a metaphor that feels complete and evocative of the human spirit—chess.

4 The Complete Metaphor: Chess and Problem Solving

The atmosphere is warm and steamy. The place, an outpost on the edge of Africa's Atlantic coast. The people who have made their way here nurse dreams of liberty behind their brittle socializing. Slowly, the camera moves in for a close-up of the owner of the café, a nightspot where the almost eternal wait disguises itself as a party. His brow is knitted as he contemplates a world at war and the misery of the human condition. An old jazz melody plays as a counterpoint to the drama that is being enacted. Another man, full of anguish (played by Peter Lorre), has come to ask a favor of Rick (played by Humphrey Bogart), who receives the request coolly. Rick has in front of him not a calculating machine or an accountant's ledger but a chessboard. Within a scene carved in wood, Rick observes a world in miniature on which he tries out the strategies that later he will have to put into play to save Ilsa (played by Ingrid Bergman) and her husband, Victor Laszlo (played by Paul Henreid), the hero of the resistance, from Nazi persecution. This film, Casablanca (1942), has taken its place in history as a paradigm of romance and separation, of the civil fight and resistance to Nazi barbarism, of integrity and chivalry versus empty words. It is no coincidence that the movie's protagonist is presented to the viewer through chess. As Emanuel Lasker, world champion for more than twenty years, said: "On the chessboard, lies and hypocrisy do not survive long."

Why Chess?

The brain and its cognitive mental processes are the biological foundation for creating metaphors about the world and oneself. Artificial intelligence, human beings' attempt to transcend their biology, tries to enter into these scenarios to learn how they function. But there is another metaphor of the world that has its own particular landscapes, inhabitants, and laws. The brain provides the organic structure that is necessary for generating the mind, which in turn is considered a process that results from brain activity.

AI studies the nature of both brain and mind, structure and function—from the perspective of mathematical models and computer science simulations. But other models and analogies come closer to the main spirit of the title of this book—chess as a total metaphor, an activity that provides a space that is rich in possibilities for research in neuroscience and artificial intelligence (figure 4.1).

The first set of metaphors leads to the cognitive processes that underlie the three characteristics that for centuries have justified the existence of chess—chess as a game, as an art, and as a science.

As a game, it lets us pinpoint all the mental processes that are necessary to generate high-level cognitive activities. These include perception and recognition of the patterns that are locked up in the sixty-four squares of the board and its thirty-two pieces; long-term memory for remembering rules and games analyzed previously; working memory for paying attention, concentrating on the game, and efficiently evaluating positions; search strategies for calculating and analyzing variations; as well as the psychological dimension that springs from a dialogue between between two brains, two ideas, and two strategic conceptions that depend on the personality of each chess player. Alfred Binet, the French psychologist who invented intelligence quotient (IQ), was fascinated by the cognitive capacity of the chess players of his time, writing, "Si l'on pouvait regarder ce qui se passe dans la tête d'un joueur, on y verrait s'agiter tout un monde de sensations, d'images, d'idées, de mouvements et de passions, un fourmillement infini d'états de conscience" (If we could look into the head of a player, we should see there a whole world of feelings, images, ideas, emotions, and passions, an ever-changing panorama of consciousness states) (1894, p. 33).

As art, chess speaks to us of the personal decisions that are made in the course of a game. Looking at this facet of the game, the essential protagonist is the aesthetic sense rather than the capacity for calculation, which thus moves us closer to the human dimension and farther from mathematical algorithms.

Finally, chess has a science-like special attraction since it lets the player first propose hypotheses of different strategic plans that are based on the game rules and possible moves of the pieces and then refute those hypotheses after careful investigation of the different lines of play. This process is analogous to the everyday work of a scientist.

A second class of metaphors—mathematical algorithms, heuristics, and models—brings us closer to the world of computer science programs, simulations, and approximations of the brain and its cognitive processes.

Figure 4.1
Chess as a metaphor of metaphors. Within chess we find thousands of years of wisdom. Game, art, science—chess is a dialogue between two conscious brains that confront plans with opposing goals and complexities, enclosing their hopes in a metareality that transcends everyday life. What remains of the richness of human chess in the artificial intelligence counterpart?

In this chapter, a brief history of chess as a human activity is presented to offer some ideas about the nature of chess. The different aspects of chess (game, art, and science) are examined, and the cognitive processes that unfold from the perspective of each of these facets are explored. The chapter ends with an introduction to how computers and mathematical methods have been used to tackle the creation of a chess program before embarking in chapter 5 on more detailed discussions about chess programming.

A Brief History of Chess

Over the years, various speculations about the origins of chess have arisen so that legends and myths have become intermingled with the historical evidence, as is inevitable with any human undertaking. But until truth is sifted out from fiction, both will be inseparable parts of the history of chess. Does chess go back to classical Greece or ancient Egypt? To long-ago Persia? Or ancient China, India or Japan? The origins of chess are so remote that it is impossible to isolate its beginnings in a single event at a specific moment in history. Chess gradually evolved in diverse regions from Asia to Europe and passing through Africa. An examination of its beginnings can help explain the extraordinary reach of the game's multiple dimensions. From its mystical origins as a dialogue with the supernatural powers to a metaphor for war, chess passes through a period as a representation of order in the universe until it becomes the game-art-science that millions of people all over the world are passionate about and that has developed into a testing ground for the sciences of artificial intelligence and cognitive psychology. What follows is a compendium of the mythological history, or historical mythology, of chess.

Occult Dialogues

Chess as a board game is part of an ancient tradition of communication with the gods that is far removed from the ludic sense it has known for the past 1,500 years. Indeed, board games seem to have begun with a mystical character that gradually changed from astrological consultation to secular entertainment. This likely occurred around the beginning of the common era somewhere along the diffuse border between western China and Kashmir, to the north of India. According to the British biologist and sinologist Joseph Needham, board games developed in ancient China to be used for divinatory arts. Thus, during the first six centuries CE, a prototype of chess was used as an exercise to determine the equilibrium between yin and yang, the two generative elements of the universe in Chinese cosmology. To do this, pieces were thrown on a board, and when they occupied certain positions, they allowed the gods to communicate with human beings. The Chinese divinatory arts survive today in the form of the *I Ching* (*Book of Changes*). These arts consist of sixty-four hexagrams, the same number as squares on the chessboard. Although the relationship is not clear, the sixty-four squares of the chess board could trace their origin to the combination of trigrams (8 x 8) of the *I Ching*. Dice were included to determine more movements of the pieces (to communicate in more detail with the gods), and later possible movements for each type of piece were defined without the use of dice. These modifications, along with the secularization of the divinatory art into a mere pastime, might have been the chain of events that led to the creation of the first versions of chess.

The Argentine writer Jorge Luis Borges, a great enthusiast of the *I Ching*, was intrigued by this relation. In his classic poem about the nature of chess, he transmits these feelings about ancestral gods:

God moves the player and he, the piece.
What god behind God originates the scheme?

Board Games and First Traces

Games with boards and pieces have been played for at least 6,000 years. Evidence remains from both ancient Egypt and Greece of many games that have been wrongly identified as the remote precursors of modern chess. For example, a fresco has been found in the tomb of Nefertari, from 1250 BCE, where the Egyptian queen is playing a game on a board. Also in Egypt, ivory towers similar to present-day chess pieces have been found and dated to about 5,000 years ago. The Greek amphora of Exekias, displayed in Villa Giulia in Rome, portrays Achilles and Ajax playing a board game. Additionally, Italian archaeologists have found chess pieces that seem to date from the second century CE, although the chronology is doubtful. Other discoveries, such as two pieces (an elephant and an ox of about 2.5 cm in height) found in Uzbekistan, also date from the second century CE. In spite of these findings, no evidence indicates that the Egyptian, Greek, and Uzbekistani games were directly related to what we today know as chess. They simply indicate the existence of board games as a constant in different civilizations.

However, the game that is closest to present-day chess is known to have developed at the beginning of the common era in northern India, on the banks of the Ganges River. That legendary game, today unanimously recognized as the ancestor of modern chess, was referred to by the Sanskrit name of *chaturanga*. After several centuries, the game became part of Indian culture and was chronicled in the verses of its contemporary poets. The first written evidence about *chaturanga* is contained in the Sanskrit poem *Vasavadatta*, which dates from the end of the sixth century CE and was written by the poet Subandhu: "The time of the rains played its game with frogs for chessmen which, yellow and green in color, as if mottled with lac, leapt up on the garden-bed squares."

In around 630, the poet Bana wrote the poem *Harshacharita*, where he notes that "only *ashtapadas* teach the position of the *chaturanga*." The *ashtapadas* that the poet refers to are boards with sixty-four squares on which a game was played whose rules have not survived. The game included pieces such as the raja, the minister, the elephant, the horse, the chariot, and the soldier. In fact, the name of the game—*chaturanga* (four parts)—refers to an Indian military formation that uses elephants, horses, chariots, and soldiers.

In eastern India, *chaturanga* probably evolved into Korean chess and the modern *Xiang Qi*, which is popular nowadays throughout China. From western India, the game moved to Persia in the early centuries of the common era and became *shatranj*. The Arab conquest of Persia (637–651) and then north Africa, southern Italy, and the Iberian peninsula determined the geographic scope for chess's conquest of western Europe (figure 4.2).

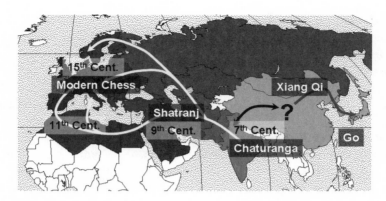

Figure 4.2
A map showing the birthplace of chess and its initial expansion.

Toward Modern Chess

The period during which the game of *shatranj* developed in the Arab world became the real foundation for modern chess. Indeed, during the blossoming of the refined, medieval Arabic culture, several books about this ancestor of chess treated the game in a systematic way and compared it to diverse mathematical structures. According to an Arabic legend about the invention of chess, the philosopher Sassa invented chess to present the Indian king Balhait with a new pastime that would distract him from his royal boredom. Pleased with this invention, the king promised the philosopher that he would give him anything he desired. The philosopher then explained to the king that he was a simple and modest man and would ask only for some grains of wheat. Nevertheless, the request should be related to the chessboard. Starting with one grain in the first square, the king had to place double the number of grains in each subsequent square. The king ordered the request be granted. If he had had any idea of modern mathematics (also developed by Arabic culture), he would have noticed immediately that the philosopher's request followed an exponential function and that the total number to fill up the board was 2^{64}, or 18,446,744,073,709,551,615 grains, enough to go around the equator billions of times.

Around the ninth century, different treatises appeared compiling positions, the first chess problems, competitions, and the names of the first chess masters. Books such as Ar-Razi's *Elegance in Chess* convey the artistic meaning that Arabic culture assigned to chess. It was also seen as a pedagogical aid for the development of logical thought, a characteristic of special interest that continues throughout history and points to the importance that chess will later have as a proving ground for the creation of artificial intelligence systems in the twentieth century.

The Arab version of *shatranj* that was transmitted to Spain (*al-shatranj* became *ajedrez*, the Spanish word for *chess*) and later to the rest of Europe greatly resembles the modern game. The most notable difference was the limited mobility of the pieces. The queen, bishops, and pawns could advance only one square with each move. For this reason, the opening consisted of a series of maneuvers that led to a point at which the pieces could begin to enter into combat. Generally, the game was much less dynamic than the one that we know. Due to the Arabic influences on the Iberian peninsula, the most important medieval book about chess was written in Spain in 1283 by Alfonso X, the Wise—*Libro del ajedrez, dados y tablas* (*Book of Chess, Dice, and Board Games*).

Nevertheless, chess spread toward Europe by many other routes, among them through Italy after the Arab conquests of Sicily and Sardinia. It also might have spread directly from Asia and Persia across the Caspian Sea toward Russia, the countries of central Europe, and even as far as the Scandinavian countries. In this way, the magic of chess would have extended throughout almost all of Europe by the eleventh or twelfth century, laying the groundwork for an intellectual enterprise without precedent in the history of board games—passing beyond the ludic borders of a pastime into the realms of art and science and ultimately offering a setting for exploring the human intellect. It is thus no wonder that Petrus Alfonsi (1062–1125) in his *Disciplina Clericalis* distinguished the following arts as necessary for the education of noble gentlemen—riding, swimming, archery, wrestling, falconry, chess, and poetry.

During the dark and devout Catholic culture of the Middle Ages, chess was one more manifestation of the order of the universe. In various writings that have survived to this day, chess was considered a metaphor of the world. Just as the pieces on the squares of the board cannot escape the player's designs, humanity cannot escape divine design (remember the Chinese origins—religion as a sublimation of astrology). Chess was soon used to teach both Christian dogma as well as morality, identifying the social order and hierarchy: the king, the queen, the advisers (bishops), the knights, the war machines, and the simple laborers. All these elements move according to rigid rules that are immutable and cannot be transgressed. The moral lesson that Christian theoreticians wanted to extract from the dynamic is clear: God disposes, and humans move only in accord with his plan, obeying the rules of the game and divine destiny.

One of the famous poems that describe this type of belief was written by Jacopo da Cessole (1250–1322). This Italian Dominican monk, who was vicar of the Genoese inquisitor for two years, delivered sermons in which each chess piece represented a different rung in medieval society. His allegorical sermons were collected in the *Book of the Customs of Men and the Duties of Nobles or the Book of Chess*. Da Cessole claims that chess was invented by the philosopher Xerxes to teach a tyrant of Babylon the error of his ways in governing. Luckily, the rigid world vision of the Middle Ages has disappeared, and gone with it are the spiritual charlatans, although nowadays we are present as a new wave of seers and critics of a rational conception of the world threaten to return us to the most recalcitrant religious obscurantism.

Although chess was seen by the church as a model of the world, one of the game's virtues—the possibility of sharing long hours with an opponent—was seen as an opportunity to pursue amorous relations. Medieval poems that portrayed a man and a woman sharing beautiful hours in front of the board (Tristan and Isolde, Lancelot and Genevieve, and, later, Ferdinand and Miranda in Shakespeare's *The Tempest*) were popular. Likewise, allegories and moral tales using chess abounded in the Middle Ages. In these long poems, each piece had a concrete meaning, and the overall game held a hidden meaning. But these allegories became so complex that the public lost interest in them. Several contemporary books have used this type of strategy and include games that have to be deciphered or that carry along the narrative thread, such as *The Squares of the City* by John Brunner or *The Flanders Panel* by Arturo Pérez-Reverte.

Modern Chess

Toward the end of the fifteenth century, chess underwent fundamental changes in its rules that turned it into the modern-day game. These changes centered on granting more mobility to the pieces. Pawns gained the opportunity to advance two squares if they were still in their original position. The bishop could make long-distance moves diagonally along the squares of its color, and the queen changed from being a weak piece to being the most powerful one on the board, uniting the movements of the rook and of both bishops. In line with these changes, in the first chess books and treatises of the fifteenth and sixteenth centuries, the new chess was given nicknames in different European languages referring to the queen's newfound protagonism—*ajedrez de la dama* (queen's chess) in Spanish, *ala rabiosa* (enraged style) in Italian, and *eschés de la dame enragée* (chess of the enraged queen) in old French.

Spain holds a place of singular historical interest in the development of chess. Continuing with the pioneering tradition begun by King Alfonso the Wise, the first book of chess published with the modern printing press was written by the Spaniard Luis Ramirez de Lucena, *Repeticion de amores e arte de axedrez* (*Repetition of the Loves and Art of Chess*), in 1497. Later, this pioneering tradition culminated in the first chess machine based on an independent algorithm, devised and created by Spanish inventor Torres y Quevedo in 1890 (see the previous chapter).

In broad strokes, the development of ideas in chess has run parallel with the evolution of the artistic currents of the past two hundred years. In fact, it is possible to speak of classicism, avant-gardism, and postmodernism in the attitudes and rules that are followed by professional players. The French musician François-André Philidor is considered the first great chess theoretician. He wrote about pawns as encompassing "the soul of chess," recognizing the fundamental character of pawn structures and establishing the first positional bases of the game. But not until a century later did the Austrian (born in Prague in 1836) Wilhelm Steinitz find the right dimension of the game, opening the way toward its systematic and rigorous study. Since Steinitz and up to the present day, the development of chess has consisted of commentaries on and rebuttals of his theories, with Emanuel Lasker, José Raúl Capablanca, Siegbert Tarrasch, and Akiba Rubinstein the most important players among those following along in the wake of the old theoretician. The hypermodern school (with Richard Reti and Aron Nimzowitsch as the main representatives), the Soviet dynamic school (with Mikhail Botvinnik, Vasily Smyslov, Tigran Petrosian, Boris Spassky, and Anatoly Karpov), and the current postmodernism that feeds off the characteristic eclecticism of our times (with Garry Kasparov as the undoubted head of that group) are all placed within the field marked out by Steinitz. Other players are difficult to categorize but have shown an astonishing capacity for the game. Alexander Alekhine, Mikhail Tal, and Bobby Fischer, all world champions, are fundamental personages in the history of chess. After establishing the bases that allowed the complexity of chess to be understood from an analytical perspective, Steinitz died alone, poor and mad, in New York in 1900.

Who Plays Chess?

Between the seventeenth and nineteenth centuries, chess consolidated itself as the intellectuals' game and moved away from its medieval obscurantism. During the years of the Enlightenment, many facets of human thought were emancipated. Art, music, literature, philosophy, and science developed at a pace that was unprecedented in the history of Western civilization. In such a historical context, chess could not remain unchanged. It developed scientifically and became deeper analytically, at the same time improving in artistic quality. Nowadays, the highly competitive character of chess attracts all kinds of people and has stopped being the exclusive patrimony of highly educated people.

Two main types of chess players are very different from each other in terms of character and attitudes toward the game. The first type of player—the *appraiser* —maintains the tradition of chess as an intellectual pursuit. This player likes to analyze positions and finds harmonious relations in them, as if she were observing a work of art. The second type of player—the *entrepreneur* (or even the *gambler*)—is attracted by the strong emotions that the game provokes by its aggressive possibilities.

The greatest chess players of all time have been a mixture of appraiser and entrepreneur, sometimes displaying more of one side or the other. Indeed, a deep understanding of the position and thinking analytically are not enough to shine in matches. Success requires that a more aggressive competitive edge be added to an intellectual stance. Emanuel Lasker, world chess champion between 1906 and 1926 and a PhD in mathematics, is an example of the intellectual type. Mikhail Botvinnik, world champion in the 1960s, also represents the analytical intellectual type. Bobby Fischer and Garry Kasparov are typical examples of the aggressive and competitive entrepreneur. A grand master's mind clearly encompasses both appraiser and entrepreneur, perhaps in parallel columns that cross in the infinity of the subconscious, where the contemplation of beauty becomes aggressive and the competitive instinct generates ideas of unexpected beauty.

The majority of chess players—those who play in neighborhood clubs and leagues and in a few tournaments—are known as *club players* and *coffee-house players*. In a local club match, it is not difficult to spot the appraiser (who is wrapped in his thoughts and always running out of his clock time) or the entrepreneur (who is a no-holds-barred gambler and strolls in to play after a sleepless night). These two types of players also roughly represent the two basic ways to play chess—the strategic or positional style versus the tactical style. Thus, the appraiser uses closed openings with delimited strategic plans, trying to acquire small advantages that accumulate until the position explodes by force toward a winning sequence. The gambler normally tries to open the position and complicate it so that the opponent's least error invites a wild sacrifice of a piece of doubtful character or a series of exchanges where the gambler has seen farther ahead and comes out the beneficiary.

Chess: A Game? A Human Problem

Chess is a game—that is, an activity to pass some free time. Its possible origins (discussed earlier) are based in its use as a medium for dialogue with the gods in ancient China, and little by little, its function in divination gave way to secular entertainment. Behind the notion of games there is a richer world that goes beyond simple pastimes. The history of the development of ludic activities in some ways is the history of civilizations. When men and women have daily interactions and common spaces are generated, activities are needed to strengthen relations among people. Actions that are vital for relating to the world also need to be practiced so that when the moment arrives to carry them out with all the consequences of real life, we are familiar with them and can react appropriately. This fascinating aspect of games generally and of chess in particular falls under the umbrella of theories of the mind. This field postulates that human consciousness evolved from a social reality in which each individual must respond to his or her own desires and the expectations of others (see the discussion of social games and chess as a metametaphor at the end of this book).

Sexual games in childhood and early adolescence are a clear example of a social game that is a preparatory activity for the moment of reproduction. Physical competitions are held in running, hunting, and swimming (activities that were fundamental to human survival in primitive societies), and board games modeled battles between two enemy armies. All these ludic activities have served the essential functions of channeling and releasing physical and psychic energies.

As a human activity, games have been subjected to scientific analysis, abstraction, and modeling. In 1944, *Theory of Games and Economic Behavior* by Oskar Morgenstern and John von Neumann opened a new chapter in the understanding of activities in which various individuals with opposing objectives interact. Morgenstern and Von Neumann's title focused on economic systems as a game in which each institution has its own objectives that in many cases directly oppose another institution's objectives. Game theory has been applied to politics, business, industrial organization, military theory, evolutionary biology, and sociology.

Game theory postulates rational behavior for each participant. Each player is conscious of the rules and behaves in accordance with them, each player has sufficient knowledge of the situation in which he or she is involved to be able to evaluate what the best option is when it comes to taking action (a move), and each player takes into account the decisions that might be made by other participants and their repercussions with respect to his or her own decision. Game theory about *zero-sum games* with two participants is relevant for chess. In this type of situation, each action that is favorable to one participant (player) is proportionally unfavorable for the opponent. Thus, the gain of one represents the loss of the other.

Game theory proposes a method called *minimization-maximization* (*mini-max*) that determines the best possibility that is available to a player by following a decision tree that minimizes the opponent's gain and maximizes the player's own. This important algorithm is the basis for generating algorithms for chess programs.

Chess: An Art? An Aesthetic Problem

Chess as a purely intellectual activity appeals to the aesthetic sense of players at all skill levels, as Richard Reti explains in his *Modern Ideas in Chess* (1974). Discovering the secrets of a position and experiencing the undoubted attractiveness of carrying out a sacrifice to secure a winning position make chess a creative activity. In the complexity of a position's labyrinth of possible variations and moves, the chess player must discern patterns and come up with ideas to carry out on the board in the same way that the artist standing in front of the paint-spotted canvas must find a harmonious solution that satisfies his creative needs. In front of the chessboard, a player is alone, silently searching to unravel the secrets of the position.

The creative possibilities in chess are more restricted than in other arts, and its movements—an opponent who insists on ruining the plans of the player, a final objective of checkmate, and a multitude of intermediate objectives (such as dominating the central squares, developing the pieces, protecting the king, securing a material advantage, and so on)—must be carried out in a limited time frame (with the exception of the open time horizon of correspondence chess).

Chess is human communication. Each player, in each move, must understand the opponent's message or soon fall into difficulties. In this way, the creative act is united with the capacity to understand the opponent's intentions, resulting in a fight of ideas, wills, and creative imagination.

This facet of chess (its nature as a fight of ideas and wills) constitutes another important aspect of chess. The will that secures a winning combination awakens in the player a sensation of achievement that is accompanied by a feeling of superiority over the rival. But if the player loses the next game, this feeling then turns into exactly the opposite as the chess player looks inside himself for the reasons for his defeat rather than looking to the opponent's skilled playing. Chess becomes an introspective window—a personal test where what matters is how the player sees himself and how he feels about his capacity to create and understand what happens on the board. And part of the act of creation during a game is the discovery of the possibilities that a position offers. Even in a phase of the game like the opening—where so much theory exists and the main variations have been studied in their minutest aspects in innumerable volumes with analyses of all the relevant possibilities—it is still possible to find something that nobody has seen before.

In *Modern Ideas in Chess*, Reti proposes two levels of recognition of the creativity and beauty in chess. The first and most evident is related to the sacrifice and its capacity to astonish, and the second is connected to an advanced knowledge of chess (such as the beauty that shines when a strategic plan is achieved). Nevertheless, sacrifice is the generator of beauty par excellence in chess. This is due to the emotions related to experiencing risk and to giving up material to secure a higher objective. There is a strain of romanticism in the sacrifice of material or "mind triumphs over matter," as Reti says. Reti has also this thought-provoking reflection: "winning combinations involving sacrifices represent to us the victory of genius over what is banal or over that jejune practical mind which seeks but to harvest every material advantage. The chess votary thus sees in the sacrifice the miraculous about which he dreams, but which as a rule he never meets with" (p. 68).

Rarely has the analogy between chess and daily life been expressed with such clarity. Both the intellectual appraiser and the enterprising gambler try to transcend the existential routine in each strategic concept and in each move. And the triumph, or failure, of an idea carries with it dimensions of what one dreams for oneself. Sacrifices in chess—the act of giving up material—is an aesthetic or even ethical proposal in the moral framework of the board, where hypocrisy does not long survive.

According to Rudolf Spielmann, a Viennese master of the early twentieth century who was known for his ability to find sacrifices even in insipid positions, there are two types of sacrifices—*temporary* and *real* ones. Temporary sacrifices imply an exact calculation of the variations. These sacrifices normally are forced and simple to calculate up to more than a dozen moves when the opponent does not have alternatives (for example, when constant checks are taking place and the opposing king must move to specific squares or otherwise perish). Real sacrifices, on the other hand, are intuitive. Calculation of the variations is not important. Instead, the player knows, intuits, or feels that giving up the material is justified and that, sooner or later, it will force a victory.

Temporary sacrifices are more frequent than real sacrifices, and although they sometimes are surprising and spectacular, they offer fewer subtleties than real sacrifices. A classic example of a temporary sacrifice, where an exact calculation of variations exists, is the Stefan Levitsky and Frank Marshall game that was played in 1912. After (1) e4 e6, (2) d4 d5, (3) ♘c3 c5, (4) ♘f3 ♘c6, (5) exd5 exd5, (6) ♗e2 ♘f6, (7) 0–0 ♗e7, (8) ♗g5 0–0, (9) dxc5 ♗e6, (10) ♘d4 ♗xc5, (11) ♘xe6 fxe6, (12) ♗g4 ♕d6, (13) ♗h3 ♖ ae8, (14) ♕d2 ♗b4, (15)

♗xf6 ♖ xf6, (16) ♖ ad1 ♕c5, (17) ♕e2 ♗xc3, (18) bxc3 ♕xc3, (19) ♖ xd5 ♞d4, (20) ♕h5 ♖ ef8, (21) ♖ e5 ♖ h6, (22) ♕g5 ♖ xh3, (23) ♖ c5 (figure 4.3), black made a sacrifice in a move known as "the most beautiful move ever played." Note how the sacrifice unites all the ingredients that are necessary to inspire a sense of beauty when contemplating it. Rowson refers to a sense of humor as characterizing these sacrifices (and chess in general), with the humor suggested by the situation of the pieces and the king (although it surely did not strike Levitsky as funny).

Real sacrifices give up material in a way that does not make clear when the decisive advantage that justifies the initial investment will be achieved. These are intuitive moves. There is no precise calculation of the variations, but the player senses that the sacrifice "must be good." A major example of real sacrifice is the game between Robert Byrne and Bobby Fischer in 1963. In this game, Fischer sacrificed a knight in move 15 without recovering material right away, leaving everyone else, including Byrne, unable to grasp the idea behind his move. Seven moves later, Byrne surrendered. Apparently, some commentators thought that Fischer had given up, given the depth of the sacrifice. The game is as follows: (1) d4 ♞f6, (2) c4 g6, (3) g3 c6, (4) ♗g2 d5, (5) cxd5 cxd5, (6) ♞c3 ♗g7, (7) e3 0–0, (8) ♞ge2 ♞c6, (9) 0–0 b6, (10) b3 ♗a6, (11) ♗a3 ♖ e8, (12) ♕d2 e5, (13) dxe5 ♞xe5, (14) ♖ fd1 ♞d3, (15) ♕c2 (figure 4.4). The following sacrifice is arguably one of the deepest in the history of chess: (15) . . . ♞xf2!!, (16) ♔xf2 ♞g4+, (17) ♔g1 ♞xe3, (18) ♕d2 ♞xg2, (19) ♔xg2 d4, (20) ♞xd4 ♗b7+, (21) ♔f1 ♕d7 0–1.

Many terms that are used to comment on games are aesthetic allusions, indicating that among chess players it is hard to separate out the game's creative and analytic aspects. Terms that are frequently used include *subtlety, depth, beauty, surprise, vision, brilliance, elegance, harmony*, and *symmetry*.

The word subtlety denotes the existence of possibilities that are hidden and not evident in an apparently stable position. One speaks then of a subtle move, like black's move 12 . . . e5 in Byrne and Fischer's game. The aesthetic pleasure that a subtle move provides is difficult to explain. All of a player's chess knowledge is wrapped up in a move that supposes the beginning of a deep plan that will culminate several moves later, revealing the original move as the key to a whole strategic idea. In the case of Fischer's move, he himself commented that he felt that, although his queen's pawn would be weakened, white would not have time to take advantage of it since the black pieces were able to take on greater mobility.

Surprise contributes to aesthetic perception and refers to moves that seem to be pulled out of a magician's hat (for example, the sacrifice of the queen in Marshall and Levistky's game). Vision is a capacity to understand a position

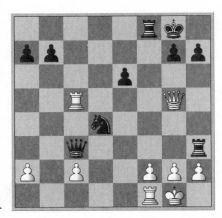

Levitsky-Marshall, 1912. Black to play.

Figure 4.3

The most beautiful chess move. Any computer chess program would find this move in less than a second. For a human player, it is a bit more complicated because the sacrifice seems to go against logical thinking. (23) . . . ♛g3!!, and white resigned. The queen cannot be taken with any of the pawns because of immediate checkmate—if (24) hxg3 ♞e2#; if (24) fxg3 ♞e2+, (25) ♔h1 ♜xf1#. A curious variation is (24) ♕e5 ♞f3+!!, putting the queen, knight, and rook en prise—(25) ♔h1 ♜xh2#. The main variation is less spectacular: (24) ♕xg3 ♞e2+, (25) ♔h1 ♞xg3+, (26) ♔g1 ♞e2+, (27) ♔h1 ♜h6, and black has an extra knight.

Byrne-Fischer, 1963. Black to play.

Figure 4.4

Bobby Fischer's sacrifice in the 1963 U.S. Championship. The winning move, (15) . . . ♞xf2!!, is so deep that when Byrne resigned, many commentators thought that Fischer had resigned instead.

and to generate solid strategic plans. And a good base of chess knowledge is needed to understand what it means to play with brilliance or elegance. The chess genius Paul Morphy is a good example of a brilliant player who always finds sharp variations, while José Raúl Capablanca is an example of an elegant player whose plans and moves are like an open book, full of clarity, logic, and energy.

Finally, harmony and symmetry are characteristics of a position that concern the positioning of the pieces and pawns on the board. When the relative positions of the pieces are in harmony, they seem to be helping each other to constitute a unified whole. Symmetry occurs between the pieces of each side and is generally maintained during some types of openings until one of the players decides to break it. Geometric motifs appear by virtue of the restrictions imposed by the squares and the rules for moving each of the pieces.

Chess also offers a modality that includes an exercise of totally free creation—*compositions*. These artificial positions are created for didactic reasons to illustrate a certain subject or to propose a problem that has to be solved following a series of indications (for example, "Black to move and mate in four"). The solution to the problem must follow the stipulation to give checkmate in four moves, neither more nor less. Some of the most beautiful positions result from compositions that surprise by their apparent simplicity and hidden subtleties that produce a feeling of astonishment and a sensation of sublime aesthetic harmony, as with the study by Richard Reti with which the book began. Here it is again (figure 4.5).

SIMPLICITY + GEOMETRY
=
BEAUTY

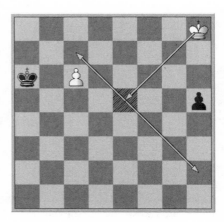

Figure 4.5
The famous ending study by Reti. White, apparently lost, moves and draws, taking advantage of the board geometry with ♔g7! Beautiful and subtle.

Adriaan De Groot claims that although the intellectual content of compositions is the same as that of the game of chess proper, few great players are at the same time problem composers. The match player has a competitive spirit that is not present in the problem enthusiast. The match player is a gambler by nature and finds great satisfaction in the hand-to-hand combat that opposes brain to brain, mind to mind, idea to idea, will to will. In contrast, the problem composer is the epitome of the appraiser, whose greatest reward is the harmonious creation of a deep composition; he is a pure artist who is absorbed in thought in the creation of beauty and the materialization of deep ideas, using the board as a canvas and the pieces as colors. An extensive literature of chess problems has its own vocabulary and technical terms where the problem composer prepares themes that develop in a different way in each problem. For example, the Grimshaw (named for its author, Walter Grimshaw, an English problem composer of the nineteenth century) is a classic theme where defending pieces with different movements (rook and bishop, for example) interfere with each other's action. When one piece moves to avoid checkmate, the defensive action of the other is blocked (figure 4.6).

GRIMSHAW THEME

Figure 4.6

Chess problems use themes with strict rules. A problemist must develop the idea (or create a new one) by placing the chess pieces in appropriate squares. Thomas Taverner composed this problem in 1881—white to move and mate in two. The key move is (1) ♖h1!, which allows checkmate by the bishop in h2, putting black in *zugzwang*. Any move to defend from checkmate with queen and knight steps into the defensive line of the other black pieces.

Chess: A Science? A Heuristic Problem

It is possible to defend the idea that any activity that allows systematic hypotheses to be generated and tested about a parcel of reality constitutes a science. The scientific method springs from observation and experimentation as bases for knowing a reality that is presented to us through the senses. Even in early childhood, this trial-and-error method takes us to a fundamental stage in our growth—the generalization of facts and the association between cause and effect. Any scientific hypothesis springs from knowledge that was previously generated by observations of facts in the real world. In addition, hypotheses produce predictions that need to be tested. For some, scientific definitions are limited to natural phenomena (although this definition would require mathematics to stop being a science since it deals with ideal objects).

But what is a natural phenomenon? If we consider for a moment, it becomes clear that every event can be classified within the realm of natural phenomena, whatever its level of complexity might be. From subatomic particles to the development of culture in the human species, everything enters the dominion of the epistemologically knowable, but each appears on its own scale of analysis. A phenomenon that exists at a complex level (such as culture) cannot be understood on the level of lesser complexity (such as brain functioning). Nevertheless, certain properties are maintained from one level of complexity to the next, which allows natural phenomena to be analyzed.

For example, the structural properties of the brain restrict the type of cognitive capacities that animals can develop, the transformations of these structural properties during evolution condition behavior, and human behaviors condition culture, and so on. Note the use of the word *conditions* instead of *determines*. Indeed, any level of lesser complexity restricts (but does not fully specify) possibilities at the higher levels where new properties can emerge. For that reason, the search for genes that determine a given behavior is futile. At the most, genes make possible (by restricting and conditioning) a behavior.

Consider the universe with its billions of galaxies, each with its billions of stars. Every civilization has been fascinated to discover one infinitesimal part of its mysteries in the form of regularities: the sun rises in the east; the moon has a cycle of different phases from full moon to new moon; comets return to pass near the earth periodically; the earth is located in one of the arms of a small spiral galaxy. We could continue enumerating numerous discoveries that fill the shelves of specialized libraries and that occasionally become general knowledge. The way that one undertakes the business of knowing the qualities of the universe is called *astronomy*, which is a scientific discipline. Does chess have sufficient qualities to be scientifically analyzed?

Chess is an activity that takes place in the perception of patterns within a closed universe, where thirty-two figures move on a fixed surface and obey precise properties or laws. Move after move, different positions form different patterns on the board. The skillful practice of chess consists in understanding these patterns as a heuristic to plan future actions and evaluate the future state of a desirable position that the player is trying to reach. The master of chess is deeply familiar with these patterns and knows very well the position that would be beneficial to reach. The rest is thinking in a logical way (calculating) about how each piece should be moved to reach the new pattern that has already taken shape in the chess player's mind. This way of facing chess is closely related to the solving of theorems in mathematics. For example, a mathematician who wishes to prove an equation needs to imagine how the terms on each side of the equal sign can be manipulated so that one is reduced to the other. The enterprise is far from easy, to judge by the more than two hundred years that have been needed to solve theorems such as that of Fermat ($z^n = x^n + y^n$), using diverse tricks to prove the equation.

The same thing happens in chess. The equation takes the following form:

$$M(P_s) = P_e$$

that is, $P_{\text{starting position}} \rightarrow P_{\text{end position}}$

where P_s and P_e represent the initial and the end board positions and the arrow M that puts them in correspondence represents the moves that are necessary to get from the starting position to the end position. The first requirement for evaluating this equation is to visualize the final position. The second requirement is to calculate moves to find a logical sequence that takes us from P_s to P_e. Neither of the two tasks is trivial. They do not even ensure that the probability that P_e is advantageous increases, due to a series of added problems, particularly the depth to which P_e has been seen. A simple example follows.

After the movements (1) e4 d6, (2) d4 ♘f6, (3) f3 g6, (4) ♗c4 ♗g7, (5) ♘e2 0–0, (6) c3 ♘bd7, (7) ♗g5 c5, (8) ♕d2 a6, (9) ♗h6 b5, (10) ♗b3 c4, (11) ♗c2 e5, (12) ♗xg7 ♔xg7, (13) g4 ♖ b8, (14) ♘g3 exd4, (15) cxd4 b4, (16) h4 ♖ e8, (17) ♔f2 ♕b6, (18) ♔g2 ♗b7, (19) h5 ♘f8, (20) hxg6 fxg6, (21) ♕g5 ♘g8, (22) ♘d2 ♘e6 (the starting position in the diagram in figure 4.7), white (the author) decided to embark on a high-risk, dubious variation with the idea of annihilating the black king's protection. After all, this is a game played with five minutes per side at the Internet Chess Club (ICC). What should be done to reach the final checkmate position shown in figure 4.7? (I did not see the

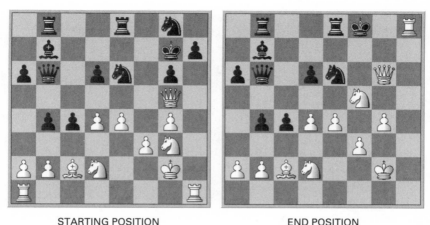

STARTING POSITION   END POSITION

Figure 4.7

Solving the Basic chess equation $\mathbf{P_s} \rightarrow \mathbf{P_e}$. White, the author, feels that the move (23) ♘f5+?! is a win. It is necessary to find a sequence of moves **M**, which makes it possible to reach the ending position $\mathbf{P_e}$ from starting position $\mathbf{P_s}$.

final position of the diagram, which in fact did not happen in the game, but my intuition made me think that the queen, knight, and rooks generated enough pressure for a decisive attack.) In the heat of the game, (23) ♘f5+?! occurred to me, a move that threatens checkmate if (23) ♔h8??.

Sometimes, reality resembles dreams, and that was indeed what happened in the game: (23) . . . ♔h8??, (24) ♖xh7+ ♔xh7, (25) ♕h4+—which was simpler than (25) ♖h1+ ♘h6, (26) ♕xh6+ ♔g8, (27) ♕xg6+ ♔f8, (28) ♖h8#, the final position shown in figure 4.7—(25) . . . ♘h6, (26) ♕xh6+ ♔g8, (27) ♕xg6+ ♔f8, (28) ♕f6+ ♔g8, (29) ♘h6+, and black resigned. Nevertheless, the simple move (23) . . . ♔f8! stops the attack, entering into a doubtful variation for white—for example, (23) . . . ♔f8, (24) ♘xc4 ♕c7, (25) ♕d2 gxf5, (26) gxf5 ♕g7+, (27) ♔f2 ♕xd4+, (28) ♕xd4 ♘xd4, (29) ♗a4 ♖e7, (30) ♘xd6 ♗c6, (31) ♗xc6 ♘xc6, and the position is equal, since white's central passed pawns compensate for the sacrificed piece.

Move Choice

The problem of choosing a move is crucial to understanding chess. Adriaan de Groot (see later in this chapter) suggested three ways to approach this question, which could be identified as decision making in chess or, as he called it, the question of freedom of choice of the player. The first is a formal approach, where the total number of possibilities is evaluated from the point of view of the legal moves; the second is the subset of objectively good moves (that is, those that meet the goal of the position, providing some type of advantage); and the third way concerns the personal decision making of the chess player (a certain psychological aspect that causes a series of moves, good and bad, to be selected). Free choice, from the point of view of the number of legal moves, ranges from 1 to more than 100 moves. Nevertheless, these extremes, especially the higher one, rarely happen. The lower end occurs in forced positions in which the king is in check and can move to only one square. De Groot did an analysis showing that around move 20, the average number of possibilities is 38, with a range from 21 to 65. The average total number of moves in a complete game is 32.5, weighing in games that reach more than 70 moves.

The problem of identifying the subset of good moves is much more complicated than simply counting the total number of possibilities and falls completely into the domain of strategy and tactics of chess as a game. Thousands of books have been dedicated to this problem, and there are still no general solutions because each position possesses particular qualities that make it unique. For a given position, there are basically two possibilities: one move might exist that is objectively better than the others, or more than one move is equally good (there is also the possibility of *zugzwang*, in which all the moves are bad). De Groot labeled the first cases as positions with an objective solution and the second cases as positions without an objective solution (in these, the number of possibilities is almost never greater than five). Examples of the first are those where there is a move that initiates a winning combination or where when, facing an opponent's check, there is only one move that defends suitably. Examples of the second often appear in openings, where it is possible to choose more than one line without one being superior to the other. De Groot comments that as openings theory advances, the number of good variations becomes more restricted, although it is also true that the opposite can happen: variations considered inferior before have been rescued from oblivion after finding a novelty.

The problem of good moves is the problem of the game of chess as a whole and provides the departure point for the development of computer programs. Finding the advantageous lines is a question of evaluating the position and searching in the tree of possible moves. If we consider the number five as a correct approximation for the subset of good moves in a given position, when calculating the number of possible terminal moves to a depth of sixteen moves (*plies*) (eight complete moves), we come close to the number of stars in our galaxy (around 100 billion, or exactly $5^{16} = 152,587,890,625$).

Nevertheless, de Groot and Gobet show that the average number of good moves considered by a grand master is 1.8 (between one and two moves before move 20). For example, during the opening, a grand master would not consider it satisfactory to play other than the following four moves—c4, e4, d4, and ♘f3—which show up in more than 98 percent of masters' games in the database of the program Junior 8. But during the opening, a player's freedom of choice is somewhat greater than in the middle game, so if we consider just two good moves to a depth of sixteen plies, the number of terminal moves to consider is reduced to 2^{16} = 62,536. That is a more reasonable number but is still impossible for any human being to calculate. Grand masters normally do not go beyond a depth of five moves (ten plies, or 2^{10} = 1,024 moves) and do so only in the main variation. But the maximum calculation when facing a complex middle-game position does not exceed five or six variations, and altogether, the exploration is less than forty or fifty moves (as an optimistic upper limit).

The third aspect in the choice of a move is related to the personality of the player. Assuming that he has been able to identify the good moves, the choice of the variation is now a question of taste or inclination toward one type of position or another. In this sense, our homemade typology of the appraiser versus the entrepreneur predicts that the first, facing equally good alternatives, will try to close the position, while the second would not hesitate to open it and launch an attack. In his well-known book *Think Like a Grandmaster*, Alexander Kotov affirms that "The direction of a player's thoughts is governed principally by the features of a given position, but no small part belongs to the character of the player. Petrosian would most likely give first thought to how to defend his weaknesses, whereas Tal would probably start to look for the chance to prepare a sacrifice" (1971 edition, p. 13). Tigran Petrosian and Mikhail Tal were world chess champions with totally opposite styles. Petrosian was known for his ultradefensive style, whereas Tal (nicknamed the magician of Riga) was capable of sacrificing pieces in the dullest of positions to complicate them. Both are extremes in mental proposals, vital attitudes in a battle of ideas that transcends the sixty-four squares of the board.

It is clear that chess constitutes an object of scientific study and has sufficient elements to be considered a science in itself. The choice of a move in chess can be analyzed as if it were a hypothesis that a player launches about the state of the position on the board. A given position shows diverse possibilities or future states that can be accessed solely by means of a series of moves (in the opening, the same position can be reached after different move sequences or *transposition*). The moves that each player makes are complementary hypotheses about the same position (the object about which the hypothesis is proposed). A chess hypothesis is basically the equivalent to drawing up a strategic plan. Experimentation in chess is equivalent to the moves that are found to carry out each plan. Throughout the history of chess, both the plans (the hypotheses) as well as the moves (the experiments) have been evolving (thanks to results from the practice of the game and from analyses), and this knowledge is the patrimony of professional players. In the reality of the match, the experiment can be tried only once, which is why the value of a hypothesis might not be easily verified. When the final position is reached, the accuracy of each hypothesis is verified or refuted. Throughout a game, each player launches numerous hypotheses about the position, elaborating strategic plans and looking for the right sequence of moves to demonstrate their validity. The player who manages to verify the last hypothesis of the game is the winner.

Chess and Cognitive Processes: First Contributions

Chess as a metaphor for understanding cognitive processes has a history of more than a century. In his book *Thought and Choice in Chess* (1965), de Groot reviews the classic contributions of chess to the analysis of different mental attributes (such as memory and attention). De Groot also addresses the work of Alfred Binet, and of Djakow, Rudik, and Petrovsky.

Psychologist Alfred Binet carried out the first study of chess as a cognitive activity, which was described in his book *Psychologie des grand calculateurs et des jouers d'échecs* (*Psychology of the Great Calculators and Chess Players*) (1894) (see chapter 2). Binet was interested mainly in the memory and visualization capacity necessary to play blindfold chess. A blindfold game is a modality of chess where at least one player, normally seated with his back to the board, plays without seeing either the board or the pieces. Playing this way has been considered since ancient times to be a prodigious demonstration of ability and intelligence, especially in exhibitions of several games played simultaneously. A very strong master is able to retain up to dozens of games in the mind and normally wins almost all of them. Many of the greatest players of all time have played renowned simultaneous matches, such as François-André Philidor at the Café de la Regence in Paris in 1744, who was carried out on people's shoulders after playing two blindfold games. Much later, the Polish-Argentine grand master Miguel Najdorf played forty-five blindfold games (although the official record is held by George Koltanowski with thirty-four games and the unofficial record by Janos Flesch with fifty-two). The great effort required for this type of activity is supposed to be dangerous for the mental health of the player, and this practice has been prohibited in various countries, such as the former Soviet Union. Neverthe-less, Koltanowski, known for holding the world record and for playing all sorts of simultaneous and blindfold exhibitions, died in 2000 at the age of ninety-six (Najdorf died at the age of eighty-seven).

Binet's study lacks rigor and is quite primitive in its methodology but is nevertheless a required reference in the history of chess and psychology. His analysis consisted of asking strong players of the time (Siegbert Tarrasch, Samuel Rosenthal, and David Janowsky, among others) to describe how they played a blindfold game. Binet's initial idea was that visual memory must be fundamental to the ability to play blindfold chess, but he reached the con-clusion that three conditions were necessary for success at blindfold chess—erudition, visualization, and memory.

For erudition, Binet established that players recognized each position on the board as a meaningful whole. Similarly, perception of the entire game was a succession of ideas and maneuvers instead of independent, uncon-nected moves. Thus, only the accumulation of experience and knowledge (that is to say, erudition) ensured that the player would have a meaningful image. The second condition, visualization (mental image), refers to the way in which the image of each position is represented in the mind of the chess player. Binet concluded that the player does not hold a complete image of the position but rather creates a global reconstruction (*gestalt*) without colors or forms. *Gestalt theory* proposes that perception and understanding constitutes an inseparable whole that is composed of all perceptible aspects (form, color, size, contrast, and so on) that are structured according to organizational laws. Finally, for the third condition, memory, Binet ended up feeling that there was no visual memory but rather some kind of abstract memory that stored a vague model of the position (including such dynamic possibilities as piece activity and threats). In this model, the memory of the position is made by means of the systematic reconstruction of the moves that preceded it. For example, Tarrasch´s report of his experience was explicit in indicating that he did not have a static image of the position as if he were seeing it but rather had to continuously reconstruct it to reach it.

As de Groot comments, the division between the condition of mental image and memory is unclear in Binet's work, possibly because he did not know how to distance himself from his first idea about visual memory. If chess were based on visual memories, players deprived of sight from birth would be incapable of learning the game, which is not the case, as is shown by the great number of masters with impaired vision. In any case, de Groot recognizes in the work of Binet a precursor to a yet little-explored area in the analysis of mental processes and indicates his admiration for the study, despite the lack of knowledge of the period and Binet's own ignorance about chess.

The second contribution to the use of chess as a cognitive laboratory was carried out in 1927 by three Soviet professors, Djakow, Rudik, and Petrovsky. These researchers invited eight players from the Moscow tournament to the psychology laboratory of the University of Moscow to participate in a series of experiments with the intention of studying what constitutes talent in chess. Some of their experiments (especially one in which the player was asked, after observing a position for a short period of time, to reconstruct it on another board) have become standard experiments in the study of cognitive capacities in chess players. Nevertheless, the work of these three researchers, who looked for correlations between chess ability and other types of mental abilities was apparently full of methodological flaws, and its results (a lack of association between chess and other abilities) have been heavily criticized.

Chess and Cognitive Processes: The Contributions of Adriaan de Groot

The classic study that linked chess, cognitive studies, and scientific experimentation is Adriaan de Groot's *Thought and Choice in Chess*. It was published for the first time in 1946 in Flemish, and the English translation appeared in 1965. In this work, de Groot set out to analyze the mental processes that underlie decision making at the moment of carrying out a chess move. To do so, he had the aid of some of the best chess players of the time, including two world champions (Alexander Alekhine and Max Euwe). In addition, he counted on the collaboration of a series of masters (today they would be considered international masters). The list of the great figures he counted on to carry out his experiments is impressive: Alexander Alekhine, Max Euwe, Reuben Fine, Salo Flohr, Paul Keres, and Savielly Tartakower, all of whom except for Tartakower participated in the famous Algemeene Vereeniging Radio Omroep (AVRO) tournament played in Holland in 1938.

De Groot's experiments were basically of two types. In the first type of experiment, a position was displayed on the board, and the player (the subject) had to think out loud about everything that was going on in his mind until he reached a decision about what move to make. The commentaries of the player were recorded, and later a series of aspects related to the game were examined. In the second class of experiment, a position was presented on the board for five seconds, and the subject had to reconstruct the original position on another board. This experiment is referred to as the standard memory task of chess. The object of this experiment was to see the types of elements that chess players recognize as piece-piece or piece-square sets. This class of analysis has led to the identification of many types of spatial relationships among pieces. In addition, the experiment has been carried out on multiple occasions to study the trajectory of the eye while a subject scrutinizes a position (thanks to special technology), which puts into correspondence the human eye's fixation space and space complexes within the chessboard. The comparative study of how players fix their perceptive attention can be used to draw conclusions about the different ways in which knowledge conditions perception and how pattern perception determines understanding.

In spite of logical reservations about the experimental protocol of the first type of experiments (since they are incapable of registering everything that happens in the mind of the chess player), the results of these experiments were of great interest. They inspired a generation of researchers to produce data of this type with chess players of different levels. Dozens of specialized research articles (some are listed in the bibliography at the end of this book) have been published, but two recent books stand out—Amatzia Avni's *The Grandmaster's Mind* (2004) and Jacob Aagaard's *Inside the Chess Mind: How Players of All Levels Think about the Game* (2004). Avni's book deciphers in an incisive and intelligent manner the way that an expert chess player thinks and offers extremely interesting ideas on the subject. In addition, it includes the thinking out loud of contemporary chess players like super-grand masters Boris Gelfand and Ilia Smirin. The second book presents readers with a series of problems that are solved out loud by players of different levels, from club players to grand masters like Artur Yusupov. Aagaard's book also shows the digital thought on the same positions from the computer software program Fritz 8.

De Groot's thinking-out-loud protocols examined the quality of the chosen movement. Since the positions were known, the main variations could be compared against the moves chosen by the participants. The length of the process until the final decision was reached also was recorded. Several other variables were recorded: the number of basal moves (defined as immediate candidates or the first move of each variation considered); the number of basal moves per minute (which provides a rate of speed for the totality of the decision process); the number of episodes (a sequence of movements that were generated by each basal move) in each variation; the number of moves in each variation; the number of nodes (positions mentioned as favorable or unfavorable after an episode) for each variation; the rate of nodes per minute for each variation; the maximum and average depth (expressed in plies or half moves) for each variation; and the duration of a first phase or orientation phase in the evaluation of the position. In the orientation phase, the player carries out a superficial examination (without calculating moves) and describes plans, threats, and number of basal moves that were reinvestigated (those that are investigated immediately after generating an episode and those that are returned to later).

These elements studied by de Groot make up the fundamental processes by which the player constructs his understanding of the position and allow the systematic study of chess thought. After studying these variables, de Groot identified three phases or subproblems that are present in all the analyses of an expert chess player's thoughts. In the *orientation phase*, the player gives a provisional version of the essential elements of the position. He is oriented to the most noticeable details of the position and begins to formulate the general problem or problems that it suggests. In the *analysis phase*, the player draws up concrete plans, mentally tests the variations, and calculates the variations to a greater depth. Finally, in the *testing phase*, the player mentally refutes the result that he has arrived at to make sure that he has not made a mistake.

The process that can be visualized in the players' thoughts is therefore a hypothetical-deductive type, beginning with the observation of the position's elements that stand out and culminating with a hypothesis such as "White has the advantage" or "I need to take advantage of the open file." These general hypotheses become more concrete hypotheses about specific moves in an iterative progression. In this clearly scientific proceeding, the player passes through a superficial phase in which she anticipates the result (feels it) based on preliminary observations. This anticipation is rarely analyzed by the less strong player but is carefully refuted by the professional player and culminates in the choice of the move.

One of the positions that is used to analyze players' thoughts is taken from one of de Groot's own games (figure 4.8). In this position, white moves and acquires a decisive advantage. For example, one of the possible winning variations—shown from the key move (1) ♗xd5 exd5, (2) ♕f3 ♔g7, (3) ♘g4 ♘xg4, (4) ♗xe7—gives white at least the exchange. Any commercial program would spit out this variation in a few seconds. The masters took more time.

The commentaries that participants in the study made are interesting. For example, Salo Flohr did not give the correct answer (although it was among the first he thought of). Paul Keres jumped directly to calculating variations, without orienting himself in the position, and in a few minutes (six) he gave the move, whereas Alexander Alekhine and Max Euwe took more time (nine and fifteen minutes, respectively, and also gave the winning move, like the rest of the masters). Alekhine gave a sample of his erudition about the position by comparing it to a game of Mikhail Botvinnik and affirming that with sufficient time he could reconstruct the complete sequence of moves

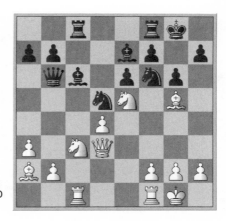

COMPLEXITY APPEARS AS AN ECHO
OF THE MULTIPLICITY OF CHOICES.

Figure 4.8

This is position A, which de Groot used in one of the tests carried out with some participants in the famous AVRO tournament in 1938 and also used later in other experiments. The best move is ♗xd5!, which takes advantage of the bad situation of the black
bishop in e7.

from the opening up to the position that was showing. Euwe made a deep
analysis of the position.

Some common denominators can be observed among the commentaries
of all the masters—an economy of ideas, a lack of depth (the variations do
not go beyond the calculation of five moves and normally they are only two),
and a final judgment, valuation, and choice of the move (with the exception of Flohr's protocol). The calculation that is necessary for arriving at a
satisfactory answer stays within the normal capacity of immediate memory
(remember the magic number of seven plus or minus two). The number
of moves considered by the five masters oscillated between three and five
(Keres considered two variations seriously, while Euwe considered five). If
each variation is counted separately, the total number of moves considered
varied from the dozen calculated by Keres to the more than thirty calculated
by Euwe. The protocols published by de Groot provide a rare opportunity for
observing the variety and common characteristics of chess players' thoughts
when faced with the same situation.

De Groot's Conclusions

De Groot reached two basic conclusions from his analyses that showed how
chess is an ideal setting for understanding varying approaches to a complex

task and that established the foundation for the cognitive psychology analyses that have been done since then.

First of all, in contrast to what had been believed, the grand masters neither calculated more variations nor went into greater depth than less strong players. Nevertheless, their answers were much better than others' answers. This pointed to the general conclusion that grand masters do not have an amazing capacity for calculating variations but rather have a large amount of chess knowledge that allows them to focus on general patterns of the position and their more relevant particular characteristics.

Second, the grand masters had a much great capacity for remembering positions than other subjects did. De Groot's interpretation, which influenced all kinds of research carried out on chess, was that masters perceive the pieces and the squares not in an isolated way but as dynamic complexes whose meaning is learned during an extensive experience of playing and study (Binet's erudition).

With these results, de Groot identified two conditions that are necessary to achieve chess mastery—an organized, meaningful perceptive episode and a collection of methods stored in memory that can be quickly accessed. In addition, he proposed the existence of two types of knowledge. *Knowing something* can be verbalized and explained, while *knowing how to do something* is intuitive and cannot be explained with words.

As Neil Charness notes, de Groot's methodological innovations (such as the protocol of thinking aloud and the experiment of recreating a position after seeing it for five seconds) have been used in the development of research protocols for cognitive psychology. Charness mentions the presentation of a pattern for a limited time to estimate expert memory in numerous groups (such as bridge players, music students, electronics technicians, basketball players, and radiologists). In this last case, for example, real x-rays were shown for 200 milliseconds (a fifth of a second) to expert radiologists, who were able to recognize the pathology in 70 percent of the cases. The majority of these experiments confirmed the correlation between the degree of knowledge and the capacity to retain a pattern.

Modern Proposals: Perception and Search

In a culture of specialization, reaching a high skill level for doing a particular kind of activity is a fundamental condition for success. The educational system itself is planned in a way that favors this specialization, sometimes to ridiculous extents, especially at the university level, where students become experts in a small subset of knowledge. Specialization is so widespread that it is worthwhile to take it as a field of inquiry on its own. Thanks to these studies, we understand the cognitive processes that are necessary to carry out a specialized activity, which also allows this understanding to be generalized to the behavior and functioning of the human mind as a whole. This type of knowledge influences educational policies, since it helps to identify the factors that are necessary for acquiring a professional specialization through learning. In this sense, chess offers an unparalleled laboratory, since both the learning process and the degree of ability obtained can be objectified and quantified, providing an excellent comparative framework on which to use rigorous analytical techniques. In addition, the clearly defined nature of the task, with its rules and objectives, and the immense database from the hundreds of thousands of master games played throughout history provide an exceptional research domain.

Since de Groot, cognitive psychologists have used chess to focus on the differences between expert and nonexpert players (with expert players understood as masters, international masters, and grand masters). The relative strength of each player is rated with an Elo score (see the appendix), which ensures a convenient distribution of strength levels that does not exist in other types of activities. For example, players with an Elo rating over 2700 (the so-called super grand masters) are clearly stronger chess players than those with ratings of 2500 (grand masters). Chess is one of the few highly skilled activities in which this kind of rating is possible. We cannot look at resumes and hospital records and easily calibrate precisely whether one surgeon is significantly better than another, for example.

Within the general framework of problem-solving theories, memory stands out as the central subject. Different models have been created to explain the way in which a player codifies the information necessary to generate a move.

The two essential components in decision making in chess are recognizing patterns stored in long-term memory (which requires an exhaustive knowledge database) and searching for a solution within the problem space. The first component uses perception and long-term memory, and the second leans mainly on the calculation of variations, which in turn has its foundations in logical reasoning. Following the meticulous analysis conducted by Fernand Gobet, a cognitive psychologist and international chess master, the next section describes two models that use these two variables to a greater or lesser extent. Generally, these models either emphasize perception elements or favor the idea that it is the capacity for calculation that prevails as a sign of mastery. The first type is represented here by the chunking theory of Simon and Chase and its improved version in the template model of Simon and Gobet; the second type, by the search model (SEEK) of Holding.

Chunking Theory

Chunking theory was initially proposed by Herbert Simon and William Chase in two influential cognitive science articles published in 1973. Their model was based on previous work by Simon and Edward Feigenbaum, in which they proposed a general theory of knowledge implemented in a program called EPAM (elementary perceiver and memorizer) that was able to carry out learning tasks. Besides the use of EPAM, chunking theory is directly connected to the findings of de Groot (described in the preceding sections).

Chunking theory indicates that experts in a given field store *chunks* (groups of related elements) of information in long-term memory that they access by means of *pointers* that are organized in a *discrimination net* that compares the perceived pattern with the data stored in long-term memory. The theory takes into account a series of known facts about working memory, such as the limitation to seven chunks (see chapter 2), the time that the discrimination net takes to recognize the pattern (roughly 10 milliseconds), and the time that a person takes to learn a new module (about 8 seconds).

In chunking theory, the chunks are like the conditions in a production system. When the discrimination net recognizes a chunk, it generates an immediate response stored in long-term memory. For example, when an open file is recognized, the immediate answer of an experienced player might be: "It needs to be occupied by a rook," which is used as a guide to find a move. This type of idea would explain the lightning speed with which a chess master can find a strong move: the different modules restrict the type of moves that are searched. In other words, the search space is immediately limited when the plan (in this case, to dominate a file) is determined and highlighted over the rest of the position's details.

In de Groot's experiments, masters' first impressions when evaluating position A included a general description of its characteristics and, in the majority of cases, an attempt to deduce the type of opening that had been used. This additional information, which a less expert player would not think of considering, already generates a set of related ideas (for example, what type of attack is used when a queen's gambit has been played). This type of immediate relation happens in different knowledge domains and helps explain what is popularly known as intuition.

The theory postulates that an expert's brain stores between 10,000 and 100,000 chunks (in other words, about the vocabulary of a language like Spanish). The English language has well over 150,000 words. An expert linguist would have between 10,000 and 20,000 words in his memory, while a normal person would have no more than approximately 10,000 words. For readers of normal texts, knowledge of some 2,000 words covers approximately 80 percent of the terms present. The remaining 20 percent of words marks the difference between knowing a language a little, having a fluid knowledge, and being an expert linguist, which in chess would be equivalent to being a club player, a national master, or a grand master. Although the calculation is approximate, if we take these figures as good approximations, the remaining 20 percent of the English language covers a grand total of at least 148,000 words.

The number of chunks necessary to reach an expert level in chess explains why ten years are needed to become a grand master (with remarkable exceptions in some child prodigies such as Samuel Reshevsky and Bobby Fischer). The thirteen-year-old Norwegian Magnus Carlsen won the Chorus C tournament of Wijk aan Zee in 2004 with an Elo rating performance of 2705. Currently, Carlsen is a super grand master and one of the strongest players in the world. In Spain, Paco Vallejo gained the title of grand master at age sixteen and currently has an Elo rating of about 2700.

Simon and Chase extended their theory with an "eye of the mind" model that explains problem-solving behavior. This model proposes the existence of a system (the eye of the mind) that holds an internal image associated with a chunk (in clear reference to mental imagery models). This system, acting as a production system, relates the information and manipulates it visuospatially in a way that allows new ideas to be abstracted from it.

In their experiments, Simon and Chase found that a piece's absolute position on the board is not as important as its relative position and capacity to carry out some action with a concrete meaning (such as control of a diagonal or file). Thus, when the subjects made a mistake, they might place a bishop in the correct diagonal but the wrong square, or they might locate a rook in the correct file but a different square from the original one. This kind of error can be explained if one assumes that the experienced chess player grants a meaning to each piece or group of pieces, so that the perception of the chunk immediately carries with it the dynamic possibilities of the position (such as mutual defense among pieces and pawns or combinatory possibilities).

An interesting experiment from Church and Church (1973) is directly related to the perception capacity of the chess player (figure 4.9). The subject is shown a board with a king of one color and a piece of another color and is asked if the king is in check. The result depends on how far the pieces are from the king: the farther away the pieces are, the more time it takes the player to perceive the relation between the attacked and the attackers. Moreover, it turns out to be much simpler to establish this relationship when the pieces are in the same rank or file than when they are in a diagonal. In addition, a classic test used to evaluate children's chess talent focuses on perception capacities. They are timed in a task that involves covering the entire board with the knight, rank by rank and square by square—from square a1 to square b1, to c1, and so on until reaching h1; from h1 to h2; and now back in the reverse direction until square a2 is reached; and so on until reaching the eighth rank. To perform this demanding task, the child needs to have a clear perception of the winding path of the knight over the chess board, a necessary quality for good players.

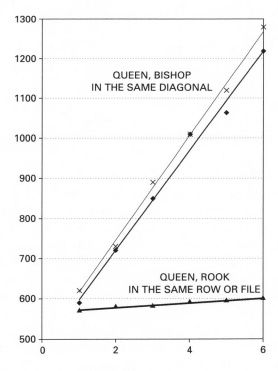

Figure 4.9
The relationship between reaction time in milliseconds (vertical axis) and distance in squares (horizontal axis) between the king and the attacking piece. The diagonal is much more complicated than files and ranks. It is more difficult to perceive that the king is in check when the queen is far away along the diagonal than when it is in the same file or rank.

The Nature of Chunks

Chess chunks are the basic vocabulary that constitutes the expert knowledge of chess players and allows them to react to a position in a matter of seconds. The initial analysis of chunks comes from experiments in reconstructing positions. The *chunk* is defined as all the pieces that are replaced with a latent period between them of less than two seconds. The pieces that are reconstructed simultaneously comprise a module that has also been perceived simultaneously. Half of the chunks turn out to be pawn structures, which is not surprising since they grant a relatively stable character to the position. They suppose an operational base that suggests a certain plan (such as occupying the weak squares and holes) and the particularities of the center (mobile, closed, open, in tension, and so on) (remember Philidor's dictum: pawns are the soul of chess). The remaining chunks are defined by relations of attack and defense, the typical castled positions, the open files and diagonals, and so on (figure 4.10). Chase and Simon list the following chunks:

• Pawns in common formations of kingside castling, possibly along with the rook and minor pieces,
• Common formations of the first rank (rooks, queen, and king),
• Pawn chains and doubled rooks and queen, and
• Attacking pieces, especially in a file or diagonal or around the king's castled position.

Experienced players retain more chunks than they theoretically could retain in working memory. Whereas experiments that interfere with the phonological loop (for example, making the subject speak) do not generate recovery problems when it comes to remembering a position set out for seconds, interference with the central executive and, above all, with the visuospatial sketchpad, however, do generate an appreciable decrease in this cognitive task.

The fact that strong players are able to retain more chunks has led to the reconsideration of the basic postulates of this theory and given rise to templates theory (see below). In addition, it has been postulated that an expert can store up to 300,000 chunks in long-term memory. Attempts have also been made to totally refute chunking theory by granting more importance to calculating variations than to the perceptive recognition of a chess pattern.

SEEK Theory

The search theory called SEEK (search, evaluation, knowledge) was proposed by Dennis Holding in the 1980s as a reaction against chunking theory. This

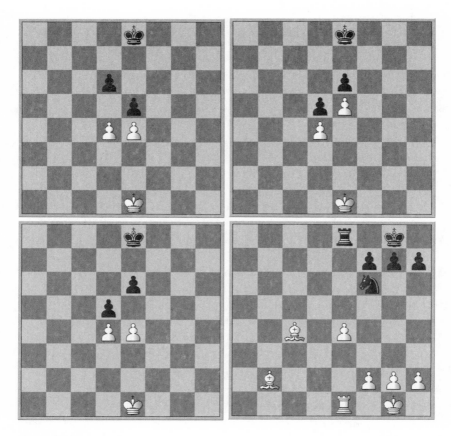

Figure 4.10

Examples of chess chunks. Central pawns form relationships that are considered classic chunks. They immediately generate several notions that are related to openings and attacking strategies. *Lower right:* A more complicated chunk with attack and defense relationships between the white bishops and the castled position of the black king. Other pieces have been omitted for clarity.

author proposed a theory that was very much in agreement with common sense: chess masters play better because they are able to search more and better, because they evaluate positions to a greater extent, and because they have a greater knowledge of chess as an activity. Holding's hypothesis maintains that of the three parameters of search, evaluation, and knowledge, search (or the capacity for calculation) is most important.

Although Holding's work has tried to refute the thesis of chunking theory, especially the importance of perception and the existence of the chunks themselves as a fundamental key to understanding expert behavior in chess players, the evidence does not seem to be in agreement with those refutations. In particular, there are studies of grandmasters' level of play during simultaneous matches, where in spite of time restrictions they are able to generate moves of great quality, which contradicts the idea that a deep search is fundamental. Thus, Gobet and Simon analyzed nine simultaneous matches played by Kasparov between 1985 and 1992 against national teams from different countries (mostly formed by an average of somewhat more than six masters of diverse categories, with an average Elo of 2413 points). Under these conditions, Kasparov had on average one-sixth of the normal time to execute a movement (if normal conditions allowed three minutes per move, then Kasparov had to think at a rate of 30 seconds per move). Despite this, the world champion had a rating performance of 2618, only about 130 points below his habitual Elo rating throughout those years.

New Ideas for Chunks: Templates

We have seen how de Groot's work has formed the basis for the development of models in chess mastery. His book, published in the 1940s, served as the basis for the first work on computer chess, published by Claude Shannon (see the following chapter). Simon and Chase's chunks theory was also built on the ideas of de Groot. In 1996, de Groot published a second book in collaboration with Fernand Gobet, *Perception and Memory in Chess*, that continues his old ideas, with the addition of the computational ideas of Gobet (de Groot himself added to the second edition of *Thought and Choice in Chess* an appendix with the rudiments necessary to create a chess program based on chess player thinking).

De Groot and Gobet published the results of extensive studies on visual fixation during the perception process of a chess position. Generally, strong players tend to fix their visual attention on the periphery of the squares that are of interest, whereas weak players concentrate on the pieces, possibly because a nonexpert player needs to think about what kind of piece he is looking at. Experts also take in more board space than weak players, which allows them to remember a greater number of details of the overall board position. In addition, expert players spend more time than nonexperts fixing their attention on empty squares. This is a fundamental difference, since the empty squares are as important as those that are occupied by pieces or pawns.

Search theory has made it necessary to modify chunks theory, especially to accommodate two issues that are highlighted by Holding's proposal—the deficiency in the size of the chunk, which is too small to clearly reflect a chess concept, and the codification time in long-term memory. Thus, chunks theory has evolved into what has come to be called *template theory*, which was elaborated by Simon himself along with Gobet. This new theory proposes that the chunks that appear time and again in the practice of chess can evolve to form larger structures, called *patterns* or *templates*, that are capable of storing an average of seventeen pieces. Templates are more dynamic than chunks in that they possess the capacity to admit variables (deviations from the standard position) that can be quickly codified in long-term memory. In other words, templates are mental schemas in which certain pieces occupy fixed places and a series of squares may or may not be occupied by pieces. Compared with chunks, this type of structure provides greater flexibility and speed when it comes to perceiving and remembering the most important factors of the position.

Chess and Personality: Psychological and Psychoanalytical Images of the Player

Vocational analysis can also study the personality of chess players. This kind of analysis considers that the choice of a type of work is related to an individual's personality type since work allows the characteristics of this personality to be expressed in an explicit way. It seems natural that other activities are also related to personal desires and that people obtain some type of reward in the form of pleasure when practicing these activities. This is the case with the majority of chess players who do not practice chess as professionals. The analysis of the personality of chess players has taken two approaches with very different philosophies. The first has looked to the psychoanalytic theories of Sigmund Freud to find aggressive repression in players, and the second has tried to develop objective studies by using personality tests to find distinguishing characteristics of players.

According to a study by Mihaly Csikszentmihalyi mentioned in an article by Avni, Kipper, and Fox (1987), although chess is a single activity, it consists of six games in one that change according to how a player prioritizes competition, problem solving, social companionship, the promotion of social status, the study of games, and relaxation. Thus, the attraction and fascination exerted by chess differ according to the type of player. There are those who enjoy the game's autotelic elements (the game as an intellectual activity) and who tend to show a low level of success. Other chess players, successful to some extent in competitive activity, tend to ignore this type of pleasure and to enjoy the competition much more (which does not mean that this type does not also enjoy the game in itself). Together with these general types, certain personality characteristics exist (particularly for autotelic players) that are associated with them—such as unconventional thinking, orderliness, or compulsive and neurotic tendencies. These three characteristics are shared by all chess players, while competitive players show characteristics like distrustfulness, aggressiveness, and hostility.

The Csikszentmihalyi study tried to decipher whether these general characteristics were accurate for a sample of sixty male individuals. Twenty men were competitive players with an average Elo of 2304, twenty men were much less competitive players with an average Elo of 1900, and twenty men were a control sample who declared that they did not play chess. Using psychological tests such as the Minnesota Multiphasic Personality Inventory, the study participants were investigated for characteristics of unconventional thought, neurosis, hostility, and distrust. Two other types of tests looked at aggressiveness and compulsiveness. The two groups of chess players turned out to be very different in the degree of unconventional thought and ordered behavior that they showed compared to the group of nonplayers. The high-level and low-level players groups were significantly different in degree of distrust, which was greater in the more competitive players. The three groups did not show any difference in aggressiveness, hostility, and neuroticism. In addition, both groups of players showed a degree of compulsive neurosis, which could be identified as a defense against external intrusions (that is, a great capacity for concentrating on a concrete problem as an excuse to isolate oneself from the outer world and thus avoid the problems of daily life). In this sense, many players commented that when they concentrated on the game, they were able to forget the passage of time and not realize what was happening around them.

Nikolai Krogius's *Psychology in Chess* (1979) is a compendium of ideas aimed at preparing players. He analyzes the images that are produced in the chess player's consciousness when a game is being played. At a general level, Krogius's analysis is similar to that of de Groot but without any experimental study. Krogius proposes the existence of three types of chess images—retained, inert, and forward. *Retained images* remain in the mind of the chess player even after he or she has reached a new position. Thus, the chess player might be sure that the blocked diagonal is still there, even though it was freed up in the previous move (see chapter 2, figure 2.3). Meanwhile, any observer can see at first glance that the diagonal is open. The problem of a chess player's retained images is a fascinating question that cannot be understood simply as a calculation error. *Inert images* in chess are those that arise in the player as a result of overconfidence in the present position, thinking that the game is over when, in reality, there is still a lot of work to be done to win. This generates a general state of relaxation, provoking errors. *Forward images* occur when the player believes that what might happen in the future is already happening in the present position. This provokes unnecessary fears and a false need to be overcautious. In addition, the forward image can induce the player into believing that his or her present position is better than it really is and to acting erroneously on that belief. The book also looks at lack of attention as a phenomenon that causes errors, the phenomenon of intuition, and the relation between the game and one of its fundamental parts, time control.

Other experiments have demonstrated that degree of emotional participation has a notable effect on solving complicated problems in chess. In these studies, participants whose galvanic response of the skin is suppressed (which is an indication of a sensitive and emotional state) displayed a greater difficulty than control subjects in solving problems. This points to something widely known: people do better with things that interest them and that feel personal because of the strength of their affection (or as an opposite but perhaps equally effective emotion, rage).

Psychoanalytic images have also been associated with chess. Some time ago, I saw an episode of the television series *Frasier* in which the protagonist buys a new chess set and challenges his father to a game. The father beats the son many times with little effort. Finally, the son wins the last game, feels guilty about having beaten his father, and at the same time feels that his father might have let him win. This slightly Freudian-hued comedy illustrates one of the ways to see the symbolism of the chess game.

Grand master Reuben Fine, one of the great players of the twentieth century, proposed a psychoanalytic theory of chess that has angered the chess community. It is the prohibited view of chess. In his efforts to build a psychoanalytic theory of the game, Fine had no recourse but to use the Oedipus complex and repressed homosexuality. His controversial book, *Psychology of the Chess Player* (1978), has become taboo for most chess players. Fine's analysis was based on investigations carried out by Ernest Jones, who saw an Oedipus complex in Paul Morphy's personality when he was unable to face the English champion Howard Staunton.

Fine postulates a Freudian analysis of chess in which the player is a repressed homosexual who plays chess to masturbate (the pieces symbolize the
penis, and rules such as "touched piece, moved piece" symbolize the social
prohibition on masturbation). At the same time, the player tries to kill the
father (the opposite king) with the support of his mother (the queen) in reference to the myth of Oedipus.

This theory, like all psychoanalytic theories, enters into lands that are difficult to verify. They move away from scientific objectivity by not being able
to be analyzed experimentally. I leave to the reader's criteria any conclusions
about the utility of these arguments.

Summary

Chess has been a metaphor from its beginnings as a dialogue between the
human and the divine to its coronation as the noblest of games. Its complexity feels overwhelming but finite, and it evokes images of what is possible and what we might like to see happen to us in real life. Each game is a
mental fight between two opposing proposals, a zero-sum game where one
player's win is the other's loss. Chess, the *Drosophila* of cognitive sciences,
has been used as a laboratory for analyzing the differences between the behavior of expert and nonexpert players. How does a player make a decision
on the best variation in a theoretical space that is as large as the number of
stars in our galaxy? Adriaan de Groot, the pioneer in the scientific analysis of
chess thinking, introduces the problematic freedom of choice that a player
faces in front of the board. We have seen that two modern theories each focus their attention on one of the two essential components of chess players'
thoughts—perception and calculation. On the one hand, the theories of
chunks and templates promote the idea of perception of chunks (perceived
complexes) as meaningful groupings of pieces and squares that the expert
player has learned throughout his career and that form a part of his long-
term memory. On the other hand, SEEK theory favors calculation capacity
as an essential element of chess strength. Now we are ready to analyze how
the silicon metaphors apply to chess.

5   Chess Metaphors: Searches and Heuristics

Edinburgh, Scotland, 1968. The Scottish chess champion, David Levy, played a game of chess with John McCarthy, creator and driving force behind the term artificial intelligence. *The atmosphere was relaxed since it was the going-away celebration of the Machine Intelligence Workshop. The champion defeated the scientist, who assured him that in ten years a computer would be able to play the revenge match for him and win it. Levy accepted the challenge in what became known as the "Levy bet." On the spot, 250 pounds sterling were put down, with 500 pounds added by Workshop organizer Donald Michie. Two years later, another programmer entered with another 250 pounds. The bet was that a computer would be able to beat Levy in a match by 1978. That year, in Toronto, Canada, David Levy played a best of six games match with CHESS 4.7, one of the strongest programs of the day, and he won his bet with two victories, two ties, and one defeat. But in a similar bet eleven years later, Levy was defeated by Deep Thought, the IBM program, by a resounding four defeats. Even in the 1978 match, the calculation power of the machines and the clever programming for evaluating positions generated good games of chess. John McCarthy congratulated David Levy in 1978 and confessed to him that he was content that a program based mainly on brute force (the calculation of variations) was not able to beat Levy. For McCarthy, the true challenge was a program based entirely on emulating the biological processes within the brain that give rise to intelligent behavior—as it is for us.*

First Programs

For artificial intelligence, a discipline that was created in the 1950s, chess is an optimal field in which to test formal languages and to elaborate specific techniques. Chess adjusted perfectly to the early expectations of AI researchers concerning the possibility that thought is computable and therefore reducible to a series of logical operations. This is the strong posture of AI. Its less optimistic position, the weak posture, feels that although a machine is able to carry out a behavior that is considered intelligent, it will never be able to understand the intelligence that it is exhibiting.

This type of analysis triggers several immediate questions: If a machine can execute the same states as a brain, is the machine thinking? Are the states of a brain system and a computer system equivalent? On what levels of the biological hierarchy can we make valid comparisons with a machine? Are the electrons that give life to a machine equivalent to the currents that are generated by sodium and potassium that travel unceasingly through the neurons of the nervous system? Scientists looked to chess as a tool for laying the theoretical bases for programs that initially could barely give a nonexpert player a decent game but that later became the chess programs that we know today and that can show up any grand master. Norbert Wiener, John von Neumann, Claude Shannon, Alan Turing, and Herbert Simon are AI pioneers, key figures in the development of science in general, who influenced the beginnings of chess programming. Other pioneers such as Charles Babbage also were interested in chess, envisioning the possibility of generating efficient algorithms to play a game. In addition, Leonardo Torres y Quevedo proposed and constructed the first autonomous machine with a built-in algorithm that could checkmate the opposite king with a rook and a king.

The fundamental problem that a chess software program must confront is the fast rate at which the possible number of plays increases (figure 5.1). In the previous chapter, it was noted that expert players do not consider more than four or five variations whenever they have to calculate a move (freedom of choice). And this is the maximum level. In addition, the depth with which each variation is examined rarely surpasses five complete moves. This supposes that, at the most, a master calculates ten moves per variation, making a total of fifty moves. All this takes place in an average time of three or four minutes per move under match conditions. On the other hand, chess programs calculate on the order of a million moves per second. That is, that in those three or four minutes, a chess program calculates on the order of 100 million moves. Even so, grand masters often beat even the best contemporary chess programs, although every year it gets more difficult.

What is missing? How it is possible that programs that have such enormous calculation capacity continue losing to grand masters? The answer rests in the quality of the moves examined. Of these moves, 99.99 percent would be considered ridiculous even by an average player. The approximate number of moves that would have to be calculated to cover the entire possible position space estimating twenty (a low estimate) as the number of possible variations to a depth of five moves gives a total of 20^{10}—that is, an order of magnitude

Figure 5.1

White's first move has to be chosen out of twenty possible moves, of which only five (including g3; see chapter 4) are considered good moves in expert chess playing because they meet the essential requirements in an opening—control of the central squares and minor piece development. To each of these twenty possible moves, black can answer with another twenty moves—that is, at move 1, the tree of possibilities is 400 moves long. The tree of moves grows exponentially: after three moves, the possibilities are on the order of 20^6, and after five moves, possibilities are on the order of 20^{10}. To avoid searching throughout this big space, algorithms and tricks are used first to find good moves and then to calculate those variations that are most promising.

greater than what modern programs calculate. Even so, although it manages to see all the variations, a depth of five moves does not absolutely guarantee that the final position will be favorable. The number of possible moves provides a factor that gives an idea of the *branchiness* of the tree of possibilities. In this case, we have given a branching factor of twenty. The larger this factor, the more difficult it will be to be able to cover all the branches of the tree. For example, in the game of Go, the branching factor is normally on the order of 100—that is, the possibility of covering the whole tree is almost null. The next sections examine, from a historical perspective, the first experiments in chess programming, and then modern programs are broken down into their constituent parts.

The First Proposal: Claude Shannon

In 1949, Claude Shannon laid out the bases for chess programming in a paper delivered at a convention on communication and information that was sponsored by the Institute of Radio Engineers (IRE). In 1950, this paper was published in the *Philosophical Magazine* and in a reduced version in *Scientific American*. Shannon's communication has a twofold historical importance for the development of computation in chess. It is the first detailed document that created a program that is capable of playing an entire game of chess with the two basic elements of any modern program—a move generator and an evaluation function for each position. And it also introduced the concept developed by John von Neumann from games theory—the *minimax algorithm*—as a strategy for calculating variations. This algorithm is the fundamental backbone of chess programs, and it has continued almost without modification to the present time. In essence, the idea of the minimax algorithm is that what is good for white is bad for black; therefore, white must look for a move that maximizes the evaluation of the position from his point of view and at the same time minimizes black's possibilities (the algorithm is described below).

The program that Shannon proposed was made up of one principal control routine and nine subroutines. Six subroutines were in charge of the rules for moving each piece, one subroutine calculated the possible variations, another reconstructed the position after a move, and the other evaluated the final position. Shannon was aware of the problem of the exponential increase of the number of moves as more variations were considered. For that reason, he showed two opposite search strategies: type A is directly subject to the consequences of exponential growth, and type B offers some solutions to this problem (figure 5.2).

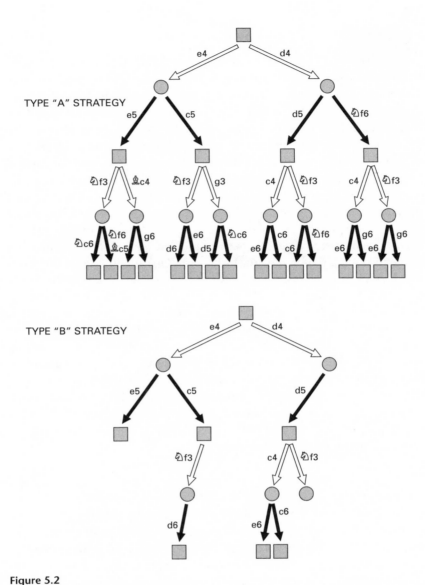

Figure 5.2
Simplified search trees. One or two variations for each move are shown (each node actually has at least twenty possible variations, with an average of thirty-five). The depth is four-ply (a drawing of a tree of depth 4 with the twenty possible moves would need to show 20^4 or 160,000 terminal nodes). Each circle represents a board position after white's move. Each arrow represents a move by white (white arrows) or by black (black arrows). For a type A strategy (brute force), all variations are analyzed up to a predefined search depth. For a type B strategy, only some variations are considered, and some of those are considered more deeply than others. Both the circles and the squares are nodes of the tree. Terminal nodes are called *leaves*. Relationships among nodes are usually known as *kin relatedness* (parental nodes, sister nodes, or descendant nodes).

Type A strategy is a brute-force strategy. It defines a depth of search before-hand and calculates all the legally possible variations. As we have seen, the astronomical number of possible moves makes this an impractical strategy because of the time that would have to be invested in analyzing variations that are absurd from the chess point of view. In addition, defining a depth of calculation beforehand means that the opportunity to analyze thoroughly the interesting variations is lost. Worse still, the program can stop analyz-ing at a critical moment in the middle of an exchange of pieces and decide that its position is better because it does not see (only because it has stopped calculating) that in the following move it loses a piece or enters a checkmate trap.

The type B strategy takes care of these problems, which is why it includes elements that allow it to discard absurd variations and to deepen variations that are evaluated as being more promising, normally in the forced varia-tions. The first important notion that arises from the problems of type A strategy is the concept of a *stable* or *quiescent position*. There are many ways to define a stable position, and from that comes the ability to understand the tactical capabilities that are hidden in the board. One way that was used by Shannon is seeing if it is possible to capture a piece in the following move. If it is, then the program continues analyzing until it is not. Still, this method can create many problems (for example, when the move following in a sup-posedly stable position is actually checkmate) (figure 5.3).

Alan Turing in Action: Turochamp

The English mathematician Alan Turing left an impressive legacy of ideas and methods for the development of artificial intelligence. The Turing test (see the previous chapter) postulated that if it is not possible to distinguish during a dialogue (of whatever kind) whether the interlocutor is a machine, then that machine possesses human behavior. Chess can provide that dia-logue. In fact, Turing analyzed the possibilities of creating a chess program based on some simple production rules—heuristics. His program is called Turochamp.

Turochamp calculated complete moves to a depth of two plies. Tur-ochamp's heuristic was based mainly on evaluating the mobility and safety of the pieces, including the king, weighing both aspects equally. For example, it evaluated mobility as the square root of the number of possible moves of a piece, and if the piece could make a capture, it counted that as an additional move. For piece security, 1 point was added if it was defended once, and 1.5 points were added if it was defended twice. In addition, pawn advances and

ONE STABLE POSITION, BUT...,
MATE IN ONE!

Figure 5.3
An example of the horizon effect. White cannot capture, so a program playing black using an algorithm evaluating the stable positions based only on capturing would reach this position and consider it stable and advantageous for black based on material difference. Since no further analyses would follow, the program would be unable to see the mate in one with the white knight.

the formation of pawn chains were favored. Finally, the heuristic gave priority to the threat of checkmate and the mere fact of putting the opposite king in check.

Turochamp used these rules (what is known as the *evaluation function*) for all the legal moves and looked deeper where it found *considerable* positions. These positions were defined as those where one could capture a piece, recover material, capture a piece defended by one of lower value (the value of the pieces was established in a more or less standard way: pawn = 1, knight = 3, bishop = 3.5, rook = 5, and queen = 10), or checkmate. If none of these conditions were fulfilled, it meant that the position was dead (similar to Shannon's quiescent or stable positions), and the function stopped calculating.

Here we see a game played in 1951 between Turing's program and a weak player who nevertheless wins the game thanks to a glaring error that Turochamp commits in move 29: (1) e4 e5, (2) ♘f3 ♘f6 (Turochamp plays the opening correctly, without knowledge of any type, merely with the simple rules of development pointed out above), (3) d4 ♗b4, (4) ♘c3 d6, (5) ♗d2 ♘c6, (6) d5 ♘d4, (7) h4 ♗g4, (8) a4 ♘xf3, (9) gxf3 ♗h5, (10) ♗b5 c6, (11) dxc6 0-0, (12) cxb7 ♖ b8, (13) ♗a6 ♕a5, (14) ♕e2 ♘d7, (15) ♖ g1 ♘c5, (16) ♖ g5 ♗g6, (17) ♗b5 ♘xb7, (18) 0-0-0 ♘c5, (19) ♗c6 ♖ fc8, (20) ♗d5 ♗xc3, (21) ♗xc3 ♕xa4, (22) ♔d2 ♘e6, (23) ♖ g4 ♘d4, (24) ♕d3 ♘b5, (25) ♗b3 ♕a6, (26) ♗c4

♗h5, (27) ♖ g3 ♕a4, (28) ♗xb5 ♕xb5, (29) ♕xd6?? (it loses immediately; the program did not calculate beyond taking the pawn in d6 because it was un-defended, consequently the resulting position did not enter within the four categories described as considerable), (29) . . . ♖ d8 0–1. Not bad for a simulation by hand (figure 5.4).

Herbert Simon

Herbert Simon is another multifaceted researcher who dedicated his career to generating ideas and models about nature. His scientific activities earned him the Nobel Prize in economics in 1978 for his contributions to the theory of decision making in financial and economic organizations. He also was interested in the keys to the functioning of the mind in problem resolution. Many of his ideas were based on the concept of *modularity* as a structural notion that allows the efficient function and evolution of a system. Chess, as might be expected, was one of his workhorses. Applying the ideas about modularity, NSS, the program developed by Simon along with Allen Newell and Cliff Shaw in 1958, had a series of modules with specific independent objectives along with a move generator.

The NSS concept went beyond a program to play chess, and for that reason it has a special relevance for us. As Simon relates, NSS was devised and constructed to show how intelligent behavior in the framework of a complex task could be achieved. NSS was one more link in the series of programs devised

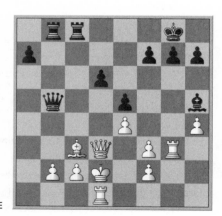

TUROCHAMP'S GAME

Figure 5.4
Position after move 28 by black. The first chess algorithm has played an average game. In the end, move (29) ♕xd6?? throws the game away.

to analyze problem solving in humans. For example, in 1960, the GPS (general problem solver) was constructed. It used an analysis strategy involving means and objectives (means and ends analysis). Given a specific objective (for example, to gain a pawn), the means and ends strategy compares recursively (over and over again) the present state of the system's situation—how many pieces are attacking and how many defending the pawn. The difference between the actual state and the objective stimulates a search in long-term memory (simulated as a database) of similar situations that are also accompanied by *operators* (concrete actions that bring the system's state closer to the given objective). For example, the gain of a pawn can be achieved by bringing in another attacking piece, exchanging a defending piece, creating another threat to attract a defending piece from another area of the board, and so on. The process is repeated recursively until it comes as close as possible to the final objective.

NSS used a selective search method that was supported by a heuristic based on operators that responded to concrete situations within the board (as in GPS) and by the attainment of objectives from specific signals. Each module was in charge of suggesting a move that was in agreement with its specific objective, which was defined according to a clear chess strategic or tactical theme—promotion of a pawn, attack on the king, control of the center, balancing material, king safety, and so on. These modules were activated only when it was necessary. For example, the module of control of the center, which suggests the move e4 or d4, is not activated during the endgame, while the module of balancing material does intervene when there are no changes possible. In this way, this program separated itself from the brute-force search strategy, calculating a much smaller number of moves in a way that was similar to how a human player calculates. NSS was the first program that used the alpha-beta algorithm as a part of the heuristic to shorten the search tree of variations carried out by the minimax (these type of algorithms are called *pruning techniques*). Together, minimax and alpha-beta (see below) form the basis of all modern chess programs.

The following game was played by NSS against Simon himself. NSS played only with the modules of material balance, control of the center, and development of the pieces. Evidently it is difficult to go beyond the opening without the more appropriate modules for the middle game and endgame, such as control of the files, diagonals, and invasion of the seventh rank. It is curious, however, to see a Nobel Prize winner in action against his own work (figure 5.5): (1) d4 ♘f6, (2) ♘c3 d5, (3) ♕d3 b6, (4) e4 ♗b7, (5) exd5 ♘xd5, (6) ♘f3 e6, (7) ♗e2 ♗e7, (8) ♗e3 0–0, (9) 0–0 ♘d7, (10) ♖ fe1 c5, (11) ♖ ad1 ♕c7, (12) ♘xd5 ♗xd5, (13) a4 ♖ ac8, (14) ♕c3 ♗f6, (15) ♗b5 ♗xf3, (16) gxf3 ♖ fd8,

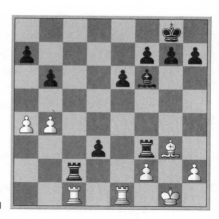

NSS'S GAME VERSUS SIMON

Figure 5.5

Position after move 26 by white. The program NSS played against one of its creators, Herbert Simon (who later became a Nobel laureate). The program played using only three modules to generate moves—central squares control, material balance, and piece development. The lack of knowledge for generating moves is evident. With a winning game, black played ♗g5 rather than d2. Even so, this move also wins.

(17) ♗xd7 ♕xd7, (18) b3 cxd4, (19) ♕d2 ♕c6, (20) ♗f4 ♕xc2, (21) ♕xc2 ♖ xc2, (22) ♖ c1 ♖ dc8, (23) ♖ cd1 ♖ 8c3, (24) b4 ♖ xf3, (25) ♗g3 d3, (26) ♖ c1 ♗g5 (the game is completely won by Simon, but (26) . . . d2! would have been more expeditious), (27) ♖ xc2 dxc2, (28) ♗e5 c1♕, (29) ♖ xc1 ♗xc1 0–1.

The NSS program was an ingenious idea that has unfortunately not been reused to generate strong chess programs, but its legacy for the later development of AI and chess programming has been incalculable. Immediately after its appearance, Alan Kotok and John McCarthy (also the creator of the programming language LISP), presented a program that used heuristics based on specific search objectives similar to those used by NSS to trim the tree. In 1966, a modified version of this program played a match against a Soviet program based on Shannon's type A strategy. The Soviet program won two games and tied two.

Brute Force or Heuristic?

In the first years of chess program development, many other proposals appeared in both the United States and the Soviet Union, and they were mainly brute-force strategies and knowledge-based strategies (heuristics to trim the search tree). The limited power of the computers available in the 1960s and

1970s did not augur a great future for the brute-force method. The search tree was so large that even a modest increase in the number of moves required a substantial increase in calculation speed. This led many researchers to criticize the brute-force method as inoperative (and, in passing, as not simulating a chess player's thoughts).

However, history has vindicated the refined brute-force method (that is, Shannon's type B strategy), using exhaustive searches, methods of pruning the useless ramifications of the tree, and libraries filled with openings and endgames. The appearance of BELLE supposed a first step in this unstoppable trend that has continued to the present time. Calculating monsters are now available that see everything (or almost everything) and that led to the great confrontations between Deep Thought, and later Deep Blue, against Garry Kasparov.

But at the other extreme of this fascinating history were some researchers who (like Simon) wanted to see a chess program that could attempt to capture the mental mechanisms of the great chess players. In this regard, the work of Mikhail Botvinnik possesses a singular importance for his personal efforts to understand and represent computationally his own behavior in front of a chessboard. And who if not the great Botvinnik could face a challenge of this caliber?

Botvinnik's ideas are based on the general principle that the objective of the player when carrying out a move is reduced exclusively to gaining material. With this principle in mind, each piece is part of a *trajectory* that takes it from the square that it occupies initially to another square where it will realize its objective. This first general principle already works as a heuristic to search for those moves that are involved in an exchange of pieces and, more concretely, in gaining material. This duality of piece and trajectory of attack forms the first level in Botvinnik's global idea of chess as a complex system. The second general principle is based on the concept of *fields* of influence for each attacking piece and its trajectory. These fields of influence for each piece are related to each other through a mathematical model that constitutes the third level of the system. PIONEER, Botvinnink's program, was guided by these heuristics to analyze deeper in a few variations and evaluated the position not as the sum of material plus other positional elements but as a function of the number of squares controlled by the pieces during the attack trajectory.

Unfortunately, as occurred with Simon's program, Botvinnik's program PIONEER and his efforts to compute the mechanisms of his own thoughts fell into oblivion. In both the former Soviet Union and the West, other programs with strategies of exhaustive searches were more successful. Among them was KAISSA, which was created by several Soviet programmers under

Mikhail Donskoy's direction. It was one of the strongest programs from the early 1970s and in 1974 won the first world chess tournament among computers. The KAISSA programmers provided many technical innovations for trimming the search tree—such as storing already found good plays in memory to be reused in new search cycles and using the technique of *windowing* (this narrows the search by initially assigning smaller scores to the parameters of the alpha-beta pruning technique, discussed later in this chapter).

Attempts to create exclusively heuristic programs included David Wilkins's PARADISE, which was based on highly elaborated production rules with concrete objectives that the program went searching for across the board. PARADISE could solve highly complex problems. Each production rule had a pattern and an action. After looking for all the possible conditions that would fulfill the pattern on the board, the program suggested an action that would be taken as a departure point to carry out an in-depth selective search. The program was made exclusively to solve problems of great tactical content, like the one shown in figure 5.6.

But the approach that continues to the present day was to make an exhaustive search using the minimax search algorithm with different methods to shorten it but without using heuristics. The CHESS program and its dif-

PARADISE IN ACTION

Figure 5.6
White to move and win. PARADISE solved this problem by computing only 109 moves, using chess knowledge encapsulated in 200 production rules. The winning move is 1.♕h5+!! with a queen sacrifice that opens the f file, allowing the action of the rook, trapping the black king into mate with a silent move. The move sequence is (1) ♕h5+ ♘xh5, (2) fxe6+ ♔g6, (3) ♗c2+ ♔g5, (4) ♖f5+ ♔g6, (5) ♖f6+ ♔g5, (6) ♖g6+ ♔h4, (7) ♖e4+ ♘f4, (8) ♖xf4+ ♔h5, (9) g3 ♕b8 (anything else also loses at once), (10) ♖h4#.

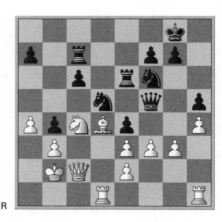

1989:
THE MACHINE BEATS THE GRANDMASTER

Figure 5.7
Position after move 26 by black. At the Long Beach tournament, in 1989, Deep Thought was first, beating very strong players. Among them, Danish grand master Bent Larsen, the first GM to lose to a machine in tournament playing conditions. With 27.g4?, Larsen enters into a combination that Deep Thought refutes easily.

ferent versions (created by a team of programmers from Northwestern University) and BELLE (from Joseph Condon and Ken Thompson) are two good examples. CHESS incorporated interesting programming subtleties to represent the information in the form of *bitboards* (which are described later). With BELLE, however, hardware specifically conceived for playing chess began to be designed, an example that many current programs follow.

All these advances, along with improvements in the architecture and computational speed of computers, contributed to gradual increases in programs' quality of play. In the 1950s, no program could beat even a weak amateur, but by the end of the 1970s, they could defeat strong players. In 1977, grand master Michael Stean lost a speed game with CHESS 4.6. But eleven years passed before a grand master lost a tournament-speed game against a machine.

This great feat was finally achieved by Deep Thought in 1989, in Long Beach, California, against Danish grand master Bent Larsen, who played white and had an Elo rating of 2560 at the time. Deep Thought was in first place in a tournament where other grand masters were participating. This game drew a spotlight of attention on the chess world. Here, Deep Thought makes an irregular defense against the English opening, and the game quickly enters into a tactics battle where the machine finally sees further than the human: (1) c4 e5, (2) g3 ♘f6, (3) ♗g2 c6, (4) ♘f3 e4, (5) ♘d4 d5, (6) cxd5 ♕xd5, (7) ♘c2 ♕h5, (8) h4 ♗f5, (9) ♘e3 ♗c5, (10) ♕b3 b6, (11) ♕a4 0–0, (12) ♘c3 b5,

(13) ♕c2 ♗xe3, (14) dxe3 ♖ e8, (15) a4 b4, (16) ♘b1 ♘bd7, (17) ♘d2 ♖ e6, (18) b3 ♖ d8, (19) ♗b2 ♗g6, (20) ♘c4 ♘d5, (21) 0–0–0 ♘7f6, (22) ♗h3 ♗f5, (23) ♗xf5 ♕xf5, (24) f3 h5, (25) ♗d4 ♖ d7, (26) ♔b2 ♖ c7, (27) g4? hxg4, (28) ♖ hg1 c5, (29) fxg4 ♘xg4, (30) ♗xg7 ♖ g6, (31) ♕d2 ♖ d7, (32) ♖ xg4 ♖ xg4, (33) ♘e5 ♘xe3, (34) ♕xd7 ♘xd1, (35) ♕xd1 ♖ g3, (36) ♕d6 ♔xg7, (37) ♘d7 ♖ e3, (38) ♕h2 ♔h7, (39) ♘f8 ♔h8, (40) h5 ♕d5, (41) ♘g6 fxg6, (42) hxg6 ♔g7, (43) ♕h7 ♔f6. 0–1 (figure 5.7).

The Structure of a Chess Program

The above sections have described the first steps taken in the history of chess programs; here the chapter turns to the ins and outs of modern programs. Computer programming for generating a chess game is a conceptually simple process that does not demand many programming subtleties regarding the three pillars on which it is based—representation, search for moves, and position evaluation. *Representation* refers to the way in which the board, the pieces, and the movements of the pieces are described numerically so they can be used appropriately by the search and evaluation algorithms. The algorithm that searches for an adequate move is supported by a *move generator* that retains a list of all the legal movements. This list is explored by the minimax to obtain a move tree (the nodes of the tree) with a defined search depth. The *evaluation* module is called on continuously by the search algorithm to pass on the scores of each position to the nodes found. Nevertheless, the devil is in the details, and although the concept and the main skeleton may be technically simple, the development of a good program with algorithms that help to trim the ramification factor and search time is an extremely complex task.

Input and Output Elements

Any computer program that depends on interaction with a user needs a way to communicate with that user. At the moment, most programs that play chess do so through a graphical interface where the board and pieces are represented. In the beginning, computers communicated with users by means of punch cards with codified messages. Monitors did not exist, and the answer was received by means of a printer. Later, moves were communicated to the program by means of a keyboard, following some type of algebraic annotation. For example, e2–e4 means "move the piece that is in square e2 to square e4." And the machine's answer, which appeared on the screen, was also given by means of algebraic notation. When graphics and the mouse appeared, the interface was finally made more natural (for programs that work on a computer).

Nevertheless, chess machines that did not depend on a computer also began to be made. These consisted of a board and a console with a small keyboard from which a move could be introduced. The move could go directly onto a board that had magnetic devices that allowed the move to be recognized when a square was pressed with a piece. A small luminous signal lit up in the square of origin and the destination square in such a way that the user did not have to do more than move the piece as indicated by the program. Although for a while the competition between programs and chess machines leaned in favor of the machines, nowadays, thanks to the popularity, speed, and storage capacity of personal computers and intuitive graphics and sound interfaces, chess programs for computers are more popular and also stronger.

Other optional elements that form a part of almost all the current chess programs include a double clock for time controls, a window with the list of moves that are generated throughout the game, and a window with the variations that the program is considering at every moment. The player makes a move using the mouse, simply dragging the desired piece from the origin square to the destination square. Once the movement is carried out, the program will automatically make its own move, and so on. As a curiosity, in 2003 there was a virtual chess experience in three dimensions in a match between Garry Kasparov and x3D Fritz. The user interface was by means of a virtual-reality projection, so Kasparov had to wear special glasses that projected the computer's image in three dimensions. Moves were introduced with a joystick.

Information Representation

The program must in some way represent everything that happens in the game. To do this, it needs a way to store in memory the state of each one of the board's squares. In other words, it requires a data structure that stores whether each of the squares is occupied or not and if it is, by which piece. In addition, it is necessary to keep track of whether castling is allowed or if it has been done already. All this will allow the move generator to know what legal moves there are in the position; to do that, this module evaluates the position moment by moment as well as after each move tried by the search algorithm. The simplest way to store the data is by means of a structure that accepts an 8 x 8 matrix where each element of the matrix is a square, with its value indicating if it is occupied or not and by which piece (figure 5.8).

```
struct position p
   {
   int board [8][8]; // use other dimensions, e.g., [10][12]
   //add sentences to check for castlings
   //add sentences to check for draw
   }
```

Figure 5.8
Data structure to store board position information.

For example, if we defined a matrix of position P[8] [8], the element of the matrix P[1][1] = 4 indicates that the square a1 has a white rook, while the element P[8] [1] = -4 indicates that the square a8 has a black rook. The values that indicate which piece is in a given square are usually (for white) 4 = rook, 2 = knight, 3 = bishop, 5 = queen, and 6 = king. Black is represented by the same values but negative. An element with value zero represents an empty square. Instead of a matrix with 8 x 8 elements, a larger matrix can be defined that has, for example, 10 x 12. With this trick, it is easier to evaluate whether a piece's move takes it off the board. These *virtual squares* are assigned different values (for example, -99) from the squares that are occupied by pieces or pawns or are empty real squares within the board, so that the program can distinguish where the eight ranks and files of the board begin and end (figure 5.9).

To determine the possibilities for each piece's move, it is necessary to verify the state of the squares to which it could theoretically accede—whether they are empty, occupied by a piece of the same color or the opposite, and able to be captured. The problem with the matrix representation 8 x 8 or 10 x 12 is that obtaining this type of information requires a large number of numerical operations that slow down the process of simply looking for possible moves.

By using so-called *bitboards*, representation operations can be made surprisingly more agile. Bitboards provide an economical representation and, at the same time, offer the possibility of executing Boolean operations on them to determine useful information about the position. A bitboard is a one-dimensional vector of sixty-four bits where each bit represents the state of a square of the board—1 if the square is occupied and 0 if the square is empty. Computer information is coded in eight-bit bytes. That means that multiples of eight are ideal for conducting operations. Although sixty-four-bit processors are already old news, the architecture of personal computers for the moment handles thirty-two bits or four bytes.

-4	-2	-3	-5	-6	-3	-2	-4
-1	-1	-1	-1	-1	-1	-1	-1
0	0	0	0	0	0	0	0
0	0	0	0	0	0	0	0
0	0	0	0	0	0	0	0
0	0	0	0	0	0	0	0
1	1	1	1	1	1	1	1
4	2	3	5	6	3	2	4

A SIMPLE 8X8 MATRIX

Figure 5.9
Board representation in an 8 x 8 matrix. Usually, the matrix is enlarged to make it easier to calculate when a piece move is illegal because it takes it outside the board.

Bitboards are a great match for sixty-four bit architectures since each one can be treated like a word by the computer, making operations on them very fast. This simple representation is powerful, not only because a single variable allows a specific characteristic of the position to be known but also because the combination among them allows numerous characteristics to be known with a minimum amount of operations, especially for the problem of legal movements. Typically, a number of bitboards are predefined to store the representation of different position characteristics. For example, there is the board of "all the pieces" (with 0s where there is no piece and 1s where there is a piece of either color), the board of "all the pawns," the board of "white pawns," the board of "black pawns," and so on.

For example, to know whether a pawn in a given square is a passed pawn, a bitboard with a 1 in the position corresponding to d5 and 0 in the others can be predefined as "passed pawn in d5." Now, a Boolean sum AND with the "black pawns" bitboard will give the answer, since it will give another bitboard with a value of 1 in the positions where there are black pawns and the white pawn in d5. If we had to conduct this operation with the traditional matrices, it would be necessary to look one after another at the value of the squares c6, c7, d6, d7, e6, and e7. Figure 5.10 shows the representation and Boolean operation AND, supposing that the black pawns are in a7, b7, f7, g7, and h7.

The bitboard produced by the operation AND has the d5 pawn isolated, which is why the return value of the operation to know whether it is isolated

```
pawn_in_d5 [64]={ 00000000 00000000 00000000 00000000
00010000 00000000 00000000 00000000 };

black_pawns [64]={ 00000000 00000000 00000000 00000000
00000000 00000000 11000111 00000000 };

pawn_in_d5 AND black_pawns [64]= { 00000000 00000000
00000000 00000000 00010000 00000000 11000111 00000000 };
```

Figure 5.10
Bitboard matrix representation and Boolean operation AND. Black pawns are in a7, b7, f7, g7, and h7. The result is an isolated d5 pawn (compare with figure 5.11).

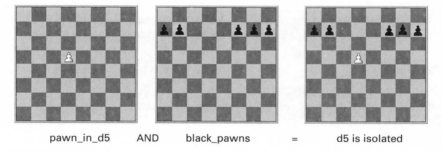

pawn_in_d5   AND   black_pawns   =   d5 is isolated

Figure 5.11
Codification of the information content of a position with bitboards. Efficient operations between bitboards are easy and give answers to complex questions very quickly. In this example, a simple AND operation can tell whether the d5 pawn is isolated or not.

would be true. Figure 5.11 shows the positional information encapsulated by the bitboards.

Representing the position by means of bitboards also allows the possibilities of attack by pawns and pieces to be quickly estimated. The knight, the pawns, and the king are easy to calculate by storing a bitboard for each attack position in all the possible squares. For example, as shown in figure 5.12, the bitboard "knight attack in d5" would give the value of 1 to the squares b4, b6, c3, c7, e3, e7, f4, and f6.

To know where the knight can attack, it is not necessary to generate more than an AND operation between this bitboard and the one that contains all the pieces of the opposite side. Only the bits that are on both boards will light

```
Knight_attack_in_d5 [64]={ 00000000 00000000
00101000 01000100 00000000 01000100 00101000 00000000 };
```

Figure 5.12
The bitboard "knight attack in d5."

up (value of 1). A trick based on *rotating* the bitboard is used to facilitate the calculation of attacking moves by the queen, bishops, and rooks.

The Evaluation Function

The evaluation function is to the chess program what the knowledge to recognize a real-life fact is to the human mind. The more knowledge that one has, the easier it becomes to recognize an event of daily life, and consequently, the action that is carried out as a reaction to that event is more satisfactory. A driver who knows how to reduce the revolutions of the car's motor by reducing the gears will be able to decelerate a car that has lost its brakes on a freeway, and this knowledge might save his life. A chess player who knows what a "queenside minority attack" is will know how to recognize the situation on the board and use the strategies that are implicit to this pattern—attack the base of the enemy chain of pawns while advancing the player's own queenside pawns supported by the rook in the first rank, wearing down the opposite position with a final incursion of pieces. For an expert chess player, the minority attack constitutes a heuristic that allows her to look for moves according to goals that are known beforehand.

In chess programs, it is necessary to evaluate the position to know how to progress within the search tree. Moreover, the data structure that contains a certain position also would have to store in a variable (for example, EVAL) the score resulting from the evaluation function to be able thus to compare the different variations during the minimax execution. The EVAL score for each position will affect which variations will be rejected or which will be analyzed more deeply using some type of criterion such as alpha-beta or many others that are mentioned below.

An evaluation function has, in essence, a formula in which different chess aspects are weighed that, in theory and practice, have been shown to be important to deciding whether a position is favorable to one side or the other. In its simpler versions, the formula consists of a linear combination of factors with a specific weight. As a result of the application of this formula, the program returns EVAL, a quantity that is scaled in relation to the value of the pieces, which oscillates around 0.0. For example, an evaluation of 3.0 means that the program evaluates the position as being favorable to white (in fact, that the advantage is sufficient to win) with a difference that is equivalent to the material value of a knight (although it is not necessarily due to white having an extra knight; it could mean, for example, two pawns and the sum of small positional advantages or any other similar combination). A negative EVAL score means that the position is favorable to black. In addition, it is necessary to program an extra factor so that the program knows when it must offer or accept a tie.

To be able to calculate the formula, the program needs to analyze the position and look for each one of the criteria that make up the equation. The chess knowledge of the programmer is encapsulated in this function, which is why it determines to a great extent the strength and the capacity to generate complex strategic plans. An example of a simple evaluation function would be an equation that represents the material balance (MB), the control and the occupation of the central squares (CS), the mobility of the pieces (MP), the control of important squares (CI), the possibility of promoting pawns (PP), king safety (KS), and the threat of mate (TM) (figure 5.13).

The parameters (k, l, m, n, p, q, r) weigh the importance of each of these factors within the evaluation function. These terms are analyzed one by one for each board position, both the actual one as well as those found by the search algorithm. Whenever the algorithm arrives at a terminal node, it calls on the evaluation function, which is why the form in which the program calculates EVAL limits the calculation speed to a great extent. Within the categories that we have distinguished, there are different variables. For example, how can the concept of mobility of the pieces be evaluated numerically? With bitboards it is simple. Boards are predefined with the mobility of any piece in any position and compared with the bitboard of the position that is being evaluated. The result is read directly.

In current programs, different types of evaluation functions exist for each phase of the game (opening, middle game, and endgame). In this way, the evaluation is much more efficient and avoids having to look for doubtful variations. However, the balance of material is the most important part of any evaluation function. If the search algorithm has seen that a sacrifice will

$$\text{EVALUATE(i): } k * MB + l * CS + m * MP + n * CI + p * PP + q * KS + r * TM$$

Figure 5.13
A simple evaluation function.

lead inevitably to a winning position, a momentary loss of material will not matter. And with this, the discussion moves on to search techniques, the most dynamic part of a chess program.

Search Algorithms: Minimax and Alpha-Beta

The search space becomes immense as it goes deeper into the number of moves. That immensity and the possibility of tackling it in an intelligent way give chess its privileged place in the world of games. Shannon proposed using the minimax algorithm (minimization-maximization) as a search tool in chess in his famous article, which was mentioned at the beginning of this chapter.

The idea of minimax is based on the following assumptions. First of all, the game is considered zero sum—that is, it starts from the idea that what is good for one player is bad for the opponent. Let us take a practical example. White plays (1) e4 in the opening. As a direct consequence, white receives a clear benefit, the domination of the center (squares d5 and f5) and the opening of diagonals d1–h5 and f1–a6 for the white bishop and white queen. For black, that gain in territory and mobility for white is counterproductive; if black does not make a move that threatens white's gain or generates the same type of benefit for its own side, black would immediately find itself in disadvantage. By responding with a move like (1) . . . e5 (the most natural response), black does the same thing as white: it gains space and mobility, equaling out the position. Second, it is assumed that each player has full information about what is happening and what can happen. In other words, the gift of omniscience is granted to each player: whenever white makes a move, black is assumed to find the best answer (figure 5.14).

Under these conditions, the minimax algorithm assumes that the first player, the maximizer, will always try to choose the move that is most beneficial to him, whereas the second player, the minimizer, will try to choose the move that does the greatest damage to his opponent. The result is that the algorithm assumes that both players will try to look for their best move (figure 5.15).

The search algorithm is the backbone of a chess program. The rest of the functions, including representation and evaluation, stay immersed within the feedback loop of a minimax. As a starting point, the search algorithm looks first for all the legal moves. For each of these moves, it then searches the variations tree as it unfolds, move by move. Depth has to be predetermined in some way, so that when the algorithm reaches a terminal node with the desired depth, no other variation will be analyzed. For this reason, the minimax algorithm, although conceptually elegant, is in reality impractical when the depth of calculation is ten complete moves.

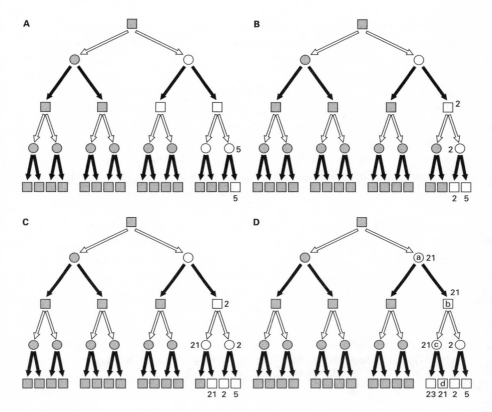

Figure 5.14

The minimax algorithm in action. The algorithm generates the tree sequentially. Whenever it arrives at a position (circles and squares in white), it calls on the module to generate all possible moves but only chooses one to analyze deeply. *A:* The algorithm has arrived at the first terminal node after black's second move and calls on the evaluation function that assigns a score of EVAL=5 to that node. *B:* Since there are no more nodes, the algorithm backs up the score to the immediately preceding node and goes down to find the second terminal node, to which it assigns a score of EVAL=2. When it returns to the preceding node, which was assigned a backed-up score of 5, the value now changes to the smaller score of 2, since this is a minimizing node. Since there are no more moves, the algorithm backs up one more node and assigns a score of 2 to the previous node. *C:* The algorithm goes down again and the process is repeated until there are no more terminal nodes in that branch. *D:* Finally, the position reached after white's first move is assigned a score of 21. That is, the program has calculated to the point where the main variation that maximizes white goes through nodes a, b, c, and d. The process repeats itself until scores are found for all terminal and intermediary nodes.

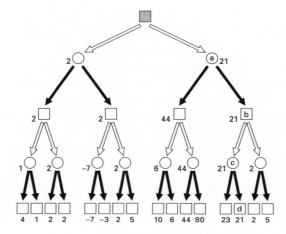

Figure 5.15
The minimax algorithm has calculated all the variations and has assigned scores to all the nodes. The a-b-c-d variation continues to be the most favorable for white, since it assigns a score of EVAL=21, no matter how well black plays.

For example, taking 35 as an average branching factor (that is, as a total number of variations) in a given position, the total number of terminal moves that would need to be calculated is 35^{20}. To reach the end of the first variation (going through all the branches), it is necessary to calculate 35^{19} terminal plays—just for the first variation. When a program like Deep Junior plays with multiple processors, it takes around one second to calculate three million moves. In other words, to go through the whole tree to a depth of only five complete moves (10 plies, 35^{10}) it would take

$$35^{10}/3 \times 10^6/60/60/24/365 = \text{more than 29 years.}$$

As is logical, the majority of these variations are useless. To reduce the tree so that the calculation of variations does not exceed twenty minutes (to a depth of five moves), the number of variations would have to be around ten or fewer, depending on the calculation capacity of the computer. Therefore, it is necessary to find some way to discard the superfluous variations without loss of information (that is, without losing the opportunity to find those that could indeed be interesting). One of the fundamental improvements to the minimax algorithm is the so-called *alpha-beta pruning technique* that was implemented in the search tree of the NSS program of Simon, Newell, and Shaw that was described above.

The alpha-beta pruning technique is complex, but the result is transparent. The algorithm picks up two additional variables when evaluating positions. It stores alpha, which represents the score of the best move for white that has been found up to the moment, and beta, with the score of black's best move. When it begins to look in another branch of the descendants of a node that shows a score smaller than alpha, the algorithm directly discards the rest of that branch, since it assumes that black will choose that variation. In other words, if it already has found a way to refute the opponent's move, there is no need to look for others. Applying the alpha-beta algorithm, the number of nodes that are visited is reduced to the square root of the initial number: for an average tree of one million nodes, a good alpha-beta can manage to reduce the number of nodes visited to 1,000 (figure 5.16).

Improving the Minimax Search Algorithm
The minimax search algorithm with alpha-beta pruning might be improved on many fronts, resulting in a total shortening of the number of terminal nodes visited, selectively improving the search of promising lines, and drastically discarding unpromising ones. Overall, what is at stake is a reduction in the time needed to come up with a good move by computationally mimicking human strategies of selective search.

A first improvement can be implemented in the programming code by means of the introduction of a simplification called *Negamax*. By means of a simple sign change, the maximizer and minimizer functions can be integrated in a single subroutine, accelerating the calculation process.

The search algorithm of *iterative deepening* constitutes another, more extensive modification of Negamax and is based on first evaluating all the variations to a depth equal to one, then to a depth equal to two, and so on, until the desired depth is reached. The great advantage of this technique is that it allows the best variation for a given time limit to be found, something that the normal minimax is incapable of doing because it can be lost trying to reach the end of one of the multiple branches when the time available to make the move has run out. If this technique is combined with a search of the main variation (that is, using the best move found in the immediately previous depth level as a first variation to analyze in the following level), then the variations are rearranged in such a way that the algorithm always looks first at the most promising variation.

Another necessary improvement is to explore variations to variable depths depending on the state of the position. For example, if the terminal node is positioned in the middle of an exchange of pieces, it does not make sense to stop calculating before the position arrives at a quiescent or calm state (this

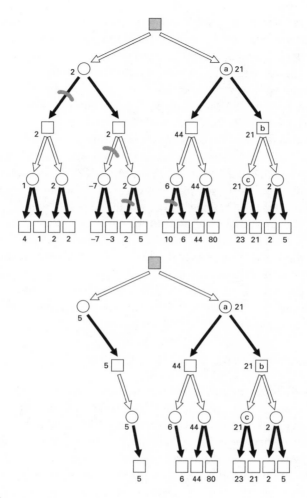

Figure 5.16

Above: Using alpha-beta pruning to make the tree much simpler. As soon as one of the scores is less favorable for white than what was found for a complete variation, there is no need to look further on that branch, since white knows that black will always try to play its best move. *Below:* The simplified search tree after pruning.

concept was present in Turing's Turochamp and in the ideas developed by Shannon). One example is the powerful search aid to minimax alpha-beta pruning that was used by Deep Blue—*singular extensions*. In this technique, a move is singular when its evaluation is much better than those of all the alternative moves. This method searches deeper for the only good move (when there is such a situation) for either side as part of the main line of play or as a move that refutes the other side's line.

There is also the use of *hash tables*, or *transposition tables*, which hold in memory positions that already were analyzed in case they occur again on another part of the tree. This happens on many occasions but above all in endgames, when there are few pieces and few moves to choose from. Related to this method is the technique of saving in memory *killer moves*—those that have already proven to be useful on another occasion (for example, because they have refuted a variation in the alpha-beta algorithm) and that are tried whenever there is an opportunity just in case they still work. In addition, the *null-move heuristic* can cut the number of branches that are explored by the minimax by looking at what would happen with the evaluation if one side theoretically skips a turn. If the evaluation is still better for that side, then that continuation is stored as a strong one.

Databases

Databases are fundamental tools for trimming the search at key moments of the game and are essential for computer programs destined to play tournaments. In classic tournament games, each player used to have two hours to think for the first forty moves. Any player who had not reached move 40 before those two hours were up lost the game. When a player reached move 40, one more hour was added. So a simple calculation of the time allowed to carry out a move in this type of game is a maximum of 120 minutes for forty moves, or an average three minutes per move.

Present-day programs incorporate *books* of openings that have been carefully compiled with the most common variations as well as those more rarely used in the practice of master-level chess. In this way, if the opponent does not depart from theory, a program can respond to the initial moves of an opening (an average of twelve moves) in a matter of a few seconds. The program then does not have to waste time in the first moves trying to find variations that have already been studied in depth by the enormous accumulated experience in the history of chess. Sometimes, these variations can continue until move 20, which means that the time per effective move doubles to six minutes. This doubling of the time that is destined to calculate each move supposes a significant increase in the number of moves that the program is able to calculate with the search algorithm, especially because this amount (six minutes) is in fact an average. Sometimes forced moves are realized in much less than a second, while in a critical moment of the game the program can take up to an hour to determine the best continuation. In any case, the fact of possessing a huge database of openings facilitates being able to carry out the first stage of the game in an automatic way. An important aspect of databases is the transposition tables that were commented on previously. These algorithms allow well-known positions to be detected even though they have been reached by a sequence of moves that were different from the opening variation that normally arrives at them. The advantage this trick grants is clear: although an initial deviation exists, the position can again be identified as belonging to a certain variation, and the program can automatically return to picking up the moves from the database. Deep Blue used to have an *extended book* (a sort of abridged version of the big database) in which several statistics of master playing, such as the number of times that a move has been played by a high-rated master, were kept in association with the big database, thus favoring good moves based on opening theory and also on the statistics of actual master play.

Books of endgames are a separate question. Chess, as a game of zero sum and total information is, theoretically, a game that can be solved. The problem is the immensity of the search tree: the total number of positions surpasses the number of atoms in our galaxy. When there are few pieces on the board, the search space is greatly reduced, and the problem becomes trivial for computers' calculation capacity. Thus, endgame databases have demonstrated how to win in different situations. Analysis of the endgame by means of algorithms takes place in a retrograde fashion—that is, from a checkmate position, the algorithm calculates how to arrive at any other position of pieces on the board. The creator of the Unix operating system, Ken Thompson, has analyzed the results of all possible endgames up to five pieces. With these databases, programs do not need to calculate variations. As soon as they detect the number of pieces, they activate the database, and the rest is automatic. On occasion, there are heuristics programmed to reach a theoretical position that results in a winning endgame for the machine. In other words, the program does not search for the win in material or in checkmate but rather for a position that it has in its database. There are distinct formats for the databases, but they are basically divided into two types. The first is based on the *distance to checkmate*. For each position, these databases store the shortest road to reaching checkmate (an example is the database of Eugene Nalimov, which is currently used by most commercial programs). The second type of endgames database is based on the *distance to conversion*. For each position, it stores the shortest road to exchanging pieces or promoting a pawn, with which the endgame changes into something simpler.

Some Results of an Academic Type

Machine-learning algorithms and bioinspired software have been gaining ground in the game arena (see chapter 3). Thus, outside of the commercial and competitive circuits, there are diverse strategies for creating chess programs that try to simulate the process of learning or the evolutionary process, sometimes combining both. In addition, there are many other interesting results that can be reached by means of different types of simulations. For example, the relationship among material, depth of search, and time is a fundamental element to understanding the dynamic relations of the pieces. Simulations with programs that face off against each other allow for the evaluation of all kinds of characteristics that shed light on different aspects of the game. In a recent work, thousands of games between different programs were generated at a depth of calculation of six plies. The conclusion drawn from this interesting experiment is that, in a game of fifty moves, an extra move advantage for one of the sides is equivalent to a minor piece (knight or bishop) and creates an advantage that leads to victory in 75 percent of cases. This is in accordance with the well known chess principle that gaining tempos and thus maintaining the initiative is sometimes worth material.

These types of results, which can be achieved only through large-scale simulations, can be used to improve the evaluation function and the efficiency of the search tree. For example, Deep Junior, reigning world champion of chess programs in 2006, considers during the minimax search that the first move of the main variation is worth two moves. This grants it much more weight than the others, and therefore, this variation ends up being explored deeper.

The strength of a program depends to a great extent on how to weigh the factors that make up the evaluation function. These factors can be changed by means of trial and error, aiming at a given objective, using what are called *evolutionary algorithms*. An entire branch of computer science applies this type of strategy to all kinds of analysis. Essentially, it simulates selective pressure, a scenario that resembles evolution by natural selection in a population of agents that can change dynamically (see chapter 3).

In chess, trial and error allows the program whose evaluation function is most successful to survive in a population of programs. Since in chess, the final objective that defines the degree of adaptation to given conditions is always the same (to checkmate the opposing king), the evolutionary problem is well defined: given an evaluation function with a series of factors that quantify the basic elements of chess analysis (material balance, control of squares, king safety, and so on), how should the parameters be weighed to change the influence of each of these factors?

There have not yet been many examples of evolutionary programming applied to chess, but its success in other fields and the boom in this new science suggests that there will be more and more experiments that pick up on this strategy of learning. For example, Graham Kendall and Glenn Whitwell made programs compete in the style of population dynamics in biology, where the best programs with the best combination of parameters values expanded their best conditions to the complete population. After forty-five generations of confrontations and mutations in the values of the parameters, the programs had advanced from an Elo score of 650 to 1750.

Evolutionary strategies of this type tend to favor the canalization of the set of values that are facing change through mutations toward values denominated *adaptive peaks*. A single evolutionary landscape can have many adaptive peaks, some higher than others, and once one of these peaks has been found, even though it is not the highest (that is, the optimal solution), it might be impossible to leave it to search for a new set of better-adapted parameters. This supposes that it is possible that following a strategy of this type (after reaching a semioptimal set of values for the parameters of the evaluation function that is located in one of these peaks), the mutations are not sufficient to make the program improve. As an additional consequence of this type of dynamic, the initial values with which the programs start the evolutionary trial are important. In the analysis of Kendall and Whitwell, who used material balance as an evaluation function, the evolutionary learning was greater when the values were fixed at the beginning as the values that chess experience grants to the pieces—1, 3, 3, 5, and 10 for pawns, knights, bishops, rooks, and queens, respectively.

The Experience with Neural Networks

Except for the evolutionary algorithms just mentioned, the chess programs that are based on the minimax algorithm with all the techniques to trim the variations tree, databases to enrich its chess knowledge, and so on have a common denominator: they do not learn from experience. Intelligent human behavior is without a doubt distinguished by its capacity for learning. From information that is incomplete, blurred, erroneous, naïve, or confusing, human intelligence is able to form criteria, recognize patterns, and make decisions about the consequences or effects of what is happening in the world. And we not infrequently get it right. To do so, the mind generates models of reality—representations of the elements that it perceives, their relationships, and the possible repercussions of those relationships in the future. How can intelligent behavior be emulated in the form of a modeling agent that is based on experience?

Besides the evolutionary strategy that was discussed in the previous section, chapter 3 introduced general ideas about neural networks as connectionist models of brain functioning. These models consist of the representation of a series of neurons that through experience learn to modulate the weight of their connections to the rest of the network's neurons. As a result, a network of this type can come to recognize guidelines within the information to which it is exposed.

In chess, there have not been many experiences of neural networks that learn to play without any type of external aid—that is, without the program's code explicitly describing the knowledge necessary to play, as with current programs. The famous checkers program of Samuel in 1959 used a learning technique by means of which the program, lacking any representation of knowledge about how to play checkers, went along learning gradually. In the 1990s, NeuroChess (from Sebastian Thrun), Morph (from Robert Levinson), SAL (from Michel Gherrity), and Octavius (from Luke Pellen) learned to play starting from zero and have reached nonexpert levels. KnightCap (from Andrew Tridgell and Jonathan Baxter) uses more encapsulated chess knowledge and has reached master level. Generally, these programs fall within the techniques of *reinforcement learning* and the majority use an algorithm of *temporal difference* learning. In essence, this computer learning paradigm approximates the future state of the system as a function of the present state. To reach that future state, it uses a neural network that changes the weight of its parameters as it learns.

Modern Human versus Machine Competitions

The work of generations—the literary dreams, the research, the technological achievements—is beginning to be lived out in our days. Chess will never go back to what it was, and modern competitions between humans and machines no longer leave any room for doubt: it will soon be impossible to beat a computer program. The table of Elo ratings for the main programs testifies to that (see the appendix). From the experience of Garry Kasparov against Deep Blue to the present time, we have witnessed an enormous improvement in chess programs' capacity for play and analysis. Below, after looking at Kasparov's experience, the 2004 tournament in Bilbao, Spain, between the best programs of the moment and three strong grandmasters—Veselin Topalov, Ruslan Ponomariov, and Sergey Karjakin—is examined.

In May 1997, at the top of the Equitable Building in Manhattan, something that many had predicted would never happen was on the verge of occurring. Massive media attention was focused on the match between Garry Kasparov and Deep Blue. This is the revenge match that followed their 1996 match in Philadelphia.

Deep Blue was the direct descendant of Deep Thought, a supercomputer with 250 chips and two processors that was created in 1988 by Feng Hsiung Hsu and Murray Campbell. In 1989, Deep Thought became the first machine able to beat a grand master (Bent Larsen) under tournament time limits. In the early 1990s, the creators of Deep Thought began to work for IBM and created Deep Blue. The system evolved to become a calculating monster that was designed to play chess with thirty-two IBM RS/6000 SP-2 nodes in parallel, containing eight VLSI chess processors as well. The Deep Blue that played Kasparov in Philadelphia possessed 256 tandem processors. The program was written in programming language C under the operating system AIX. Kasparov lost the first game but won the match 4–2.

In 1997, Deep Blue had processors that were twice as fast as the ones that it had in 1996. With a calculation speed of more than 200 million moves per second, Deep Blue was able to calculate a complete tree of ten variations to a depth of ten moves in less than ten minutes. Reducing the branching factor to five, Deep Blue could calculate, under tournament time limits, up to a depth of sixteen moves. Grand master Joel Benjamin was in charge of improving the openings book, for which IBM also hired Spanish grand master Miguel Illescas. Drawing an unexpected amount of media interest, Deep Blue won the match, winning the last game in a way that to more than one commentator seemed strange, leading to speculation about the possibility that the match was fixed. Be that as it may, here is the last game between Deep Blue and Garry Kasparov, for the first time in history giving a victory to a machine over the best chess player of his day: (1) e4 c6, (2) d4 d5, (3) ♘c3 dxe4, (4) ♘xe4 ♘d7, (5) ♘g5 ♘gf6, (6) ♗d3 e6, (7) ♘1f3 h6, (8) ♘xe6 ♕e7, (9) 0–0 fxe6, (10) ♗g6 ♔d8, (11) ♗f4 b5, (12) a4 ♗b7, (13) ♖e1 ♘d5, (14) ♗g3 ♔c8, (15) axb5 cxb5, (16) ♕d3 ♗c6, (17) ♗f5 exf5, (18) ♖xe7 ♗xe7, (19) c4, 1–0 (figure 5.17).

The match between Kasparov and Deep Blue made news around the world as the triumph of machines and the defeat of humanity. It also was a marketing operation that increased the value of IBM's shares in the stock market.

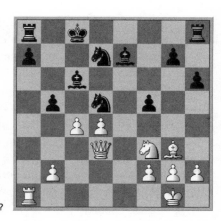

HOW IS IT POSSIBLE?

Figure 5.17

The last game of the match between Deep Blue and Kasparov. After following a theoretical Caro Kann variation in which white sacrifices a knight for a pawn and puts a lot
of pressure on the black king, Kasparov made a dubious move, (11) . . . b5?!, and the
machine won the game some moves later. Many people still can't believe it.

But what is the true meaning of this encounter? Humans have created machines that imitate us—that provide mirrors to see ourselves and measure
our strength, our intellect, and even our creativity. Deep Blue, the calculating
monster with hundreds of parallel processors, calculating million of moves
per second, would have seemed pathetic if it had not won the match. Leaving
aside the details of the match itself, Deep Blue's victory increases even more
the stature of Kasparov and the rest of us mortals, who with our emotions,
dreams, hungers, desires, hopes, and fears are able to face monsters of such
mythological proportions. The advance of science and knowledge presents
us with multiple dilemmas, among them the fear of knowing ourselves that
is intermingled with curiosity, as if we were children exploring dark rooms
by the meager light of a candle. We never know where and when exactly the
next bogeyman will jump out.

In October 2004, a tournament between chess programs (Fritz 8, Deep Junior, and Hydra) and three strong players (super grand masters Veselin Topalov and Ruslan Ponomariov and then twelve-year-old Sergey Karjakin, the
youngest grand master in the history of chess) took place in Bilbao, Spain.
The programs ran in different platforms to provide an ample range of processing speed. Fritz 8 ran on a laptop with a 1.7 GHz Centrino processor, Deep
Junior ran on a Xeon computer with four 2.8 GHz processors operated from
England, and Hydra operated as a chess platform that integrated hardware

especially designed for the program with sixteen processors located in the United Arab Emirates. The games were played with tournament time limits of two hours for forty moves and an additional one hour for the rest of the game.

The result takes us into the future. There were four days of play. The first day brought a result of 0.5 points to 2.5 points in favor of the programs (a player earns 0.5 points for a tie; the winning side earns a full point). The second day was favorable to the grand master team—2 to 1. But the third and fourth days showed the same result as the inaugural day—0.5 to 2.5.

Let us look at the following game from a former FIDE (Fédération Intér-national des Echecs, the World Chess Federation) world champion, Ruslan Ponomariov (2,710 Elo), against Hydra. This is how programs win these days: (1) ♘f3 ♘f6, (2) c4 b6, (3) d4 e6, (4) g3 ♝a6, (5) b3 ♝b4, (6) ♝d2 ♝e7, (7) ♝g2 c6, (8) ♝c3 d5, (9) ♘e5 ♘fd7, (10) ♘xd7 ♘xd7, (11) ♘d2 0–0, (12) 0–0 b5, (13) c5 e5, (14) b4 e4, (15) e3 ♕c7, (16) ♖ e1 ♝g5, (17) a4 bxa4, (18) ♖ xa4 ♝b5, (19) ♖ a3 ♘f6, (20) ♝f1 a6, (21) ♕a1 ♖ ab8, (22) ♝xb5 axb5, (23) ♖ a7 ♕c8, (24) ♕a6 ♕e6, (25) ♖ a1 h5, (26) ♖ c7 h4, (27) ♖ xc6 ♕f5, (28) ♕a2 ♘g4, (29) ♘f1 ♕f3, (30) h3 ♘xe3, (31) fxe3 ♝xe3, (32) ♔h2 ♝f2, (33) gxh4 e3, 0–1 (figure 5.18).

Hydra's game against Ponomariov is conclusive. The program overwhelms the former champion and leaves him in a position that, as Rowson says, is almost humorous. The program sacrifices material for a winning position.

SAD POSITION FOR A HUMAN

Figure 5.18
Ponomariov gave up after black's last move, (33) . . . e3. The position is desperate for white despite its extra knight. The match is against Hydra, a novel chess platform that plays with sixteen parallel processors located in the United Arab Emirates.

The border between human depth and computer materialism is beginning to disappear. This is not necessarily thanks to the calculation capacity of computers. Careful programming has meticulously coded hundreds of years of chess knowledge, and the evaluation functions are capable of analyzing with unusual depth the subtleties and the true value of positions that occur throughout the game.

Fritz won a game playing in a simple laptop against no more nor less than Topalov (Elo 2757), one of the best present-day players. Here there are no parallel processors or chips dedicated exclusively to the task of winning at chess. The game was played on a laptop slower than the one that I am using to write these lines. The course of the game leaves no doubts: (1) e4 e6, (2) d4 d5, (3) ♘d2 a6, (4) ♘gf3 ♘f6, (5) e5 ♘fd7, (6) ♗d3 c5, (7) c3 ♘c6, (8) 0–0 g5, (9) ♗b1 g4, (10) ♘e1 h5, (11) ♘b3 a5, (12) ♘xc5 ♘xc5, (13) dxc5 ♗xc5, (14) ♘d3 ♗a7, (15) ♕a4 ♗d7, (16) ♕f4 ♗b8, (17) ♖ d1 f5, (18) c4 d4, (19) ♖ e1 ♕e7, (20) ♗c2 h4, (21) ♗d2 ♗c7, (22) ♗d1 ♖ g8, (23) a3 a4, (24) f3 gxf3, (25) ♗xf3 ♗a5, (26) ♗xa5 ♖ xa5, (27) ♕d2 ♕g5, (28) ♕f2 ♖ g7, (29) c5 ♔f8, (30) ♖ ac1 ♔g8, (31) ♔h1 ♖ a8, (32) ♖ c4 ♖e8, (33) ♗d1 ♕h6, (34) ♘f4 ♖ d8, (35) ♖ xa4 d3, (36) ♗b3 ♗f7, (37) ♕e3 ♕g5, (38) ♖ d1 ♔h7, (39) ♖ d2 ♕h6, (40) ♔g1 ♕g5, (41) ♗c4 ♗e8, (42) ♗xe6 ♖ e7, (43) ♖ xd3 ♖ xd3, (44) ♕xd3 ♘xe5, (45) ♕xf5 ♕xf5, (46) ♗xf5 ♔h6, (47) ♘d5, 1–0 (figure 5.19).

In Bilbao, machines defeated humans by a resounding 8.5 to 3.5. Fritz secured 3.5 points from its four games from a simple laptop running at 1.7

FROM A LAPTOP

Figure 5.19
Topalov, with black, loses against Fritz 8. The game was especially relevant because the program ran on a simple laptop computer. In today's computer chess, calculating power coexists with search techniques and knowledge programming.

GHz. The Bilbao tournament is an instructive experience that reminds us that chess is not infinite but indicates the triumph of a human intelligence that is capable of creating machines of astounding capabilities.

In November 2006, another man-versus-machine match took place in Germany, this time between reigning world champion, Vladimir Kramnik, and Fritz 10. As if destined to show the fate of computer chess for the twenty-first century, Fritz 10 overwhelmed Kramnik with a 4–2 score.

The Internet Experience

The development of computer science inevitably passes through the Internet and the almost miraculous possibility of communication in real time between terminals that are on opposite sides of the planet. The ability to share and to consult information of all types (and quality) is also revolutionizing the development of programs and computer utilities thanks to initiatives like GNU.org (where users can access the source code of programs to improve and expand them) and multilingual encyclopedias like Wikipedia.org (an experience in information usage, where users add or perfect the contents). All this is democratizing access to and the use of data that even ten years ago would have been unthinkable to find in a reasonable amount of time. This reality has affected chess by increasing opportunities to play and access to information. In terms of playing, the horizon has opened up in such a way that now it is hardly necessary to leave the house to play, at any hour, against players all over the world of every level, class, and condition.

Here is a routine scene for thousands of Internet users around the world. It is three in the morning and too hot to sleep. I try changing position, counting backward, counting sheep. Nothing helps. I decide that it is better to get up and do something. I turn on the computer, connect to the Internet, and in less than five minutes have entered one of the numerous sites where thousands of players worldwide meet to play chess. There are pseudo Elo rating lists that group players according to their relative strength and screens to post comments. In seconds, a window opens to tell me that there is a player inviting me to a game. It is someone in my category who wants to play a fast game of five minutes. I accept, and we immediately go to another window where there is a board with the pieces already placed, a clock, and a small window in which we can send messages. We greet each other. My opponent is in Argentina, where it is 10 o'clock at night. We wish each other good luck and start the game.

Besides these experiences, which are already commonplace for hundreds of thousand of chess enthusiasts around the world, the Internet offers the possibility of novel events (see also the appendix for the addresses of the most important chess sites for playing and information). One of these could not have taken place in such a dynamic way if it were not for the Internet—Kasparov's match against "the rest of the world," which was played between June and October 1999. Kasparov, with white, played at a rhythm of forty-eight hours per move against a worldwide voting.

Four teams led by strong subseventeen players proposed moves, and the Internet observers throughout the world voted on them. The one that received the most votes was played. To get an idea of the popularity of this event, the servers received an average of 200,000 votes every day. The game finished in controversy, since on move 51 of black there were problems receiving one of the team's suggestions, indicating that a hacker had broken in to the server and forced the vote for the move that was finally made at the expense of a stronger continuation. In this way, the game entered into a winning variation for Kasparov.

This is the game of Kasparov versus the world: (1) e4 c5, (2) ♘f3 d6, (3) ♗b5+ ♗d7, (4) ♗xd7+ ♕xd7, (5) c4 ♘c6, (6) ♘c3 ♘f6, (7) 0–0 g6, (8) d4 cxd4, (9) ♘xd4 ♗g7, (10) ♘de2 ♕e6, (11) ♘d5 ♕xe4, (12) ♘c7+ ♔d7, (13) ♘xa8 ♕xc4, (14) ♘b6+ axb6, (15) ♘c3 ♖ a8, (16) a4 ♘e4, (17) ♘xe4 ♕xe4, (18) ♕b3 f5, (19) ♗g5 ♕b4, (20) ♕f7 ♗e5, (21) h3 ♖ xa4, (22) ♖ xa4 ♕xa4, (23) ♕xh7 ♗xb2, (24) ♕xg6 ♕e4, (25) ♕f7 ♗d4, (26) ♕b3 f4, (27) ♕f7 ♗e5, (28) h4 b5, (29) h5 ♕c4, (30) ♕f5+ ♕e6, (31) ♕xe6+ ♔xe6, (32) g3 fxg3, (33) fxg3 b4, (34) ♗f4 ♗d4+, (35) ♔h1 b3, (36) g4 ♔d5, (37) g5 e6, (38) h6 ♘e7, (39) ♖ d1 e5, (40) ♗e3 ♔c4, (41) ♗xd4 exd4, (42) ♔g2 b2, (43) ♔f3 ♔c3, (44) h7 ♘g6, (45) ♔e4 ♔c2, (46) ♖ h1 d3, (47) ♔f5 b1♕, (48) ♖ xb1 ♔xb1, (49) ♔xg6 d2, (50) h8♕ d1♕, (51) ♕h7 b5, (52) ♔f6+ ♔b2, (53) ♕h2+ ♔a1, (54) ♕f4 b4, (55) ♕xb4 ♕f3+, (56) ♔g7 d5, (57) ♕d4+ ♔b1, (58) g6 ♕e4, (59) ♕g1+ ♔b2, (60) ♕f2+ ♔c1, (61) ♔f6 d4, (62) g7, 1–0 (figure 5.20).

Kasparov's experience is an example of the convocational power of the Internet—its democratization of the access to and use of information and its transcending of space and time to unite communities all over the planet around common passions. Chess is one of them, but the opportunities that this new form of social interaction offers us are infinite. The ability to play online at any hour is revolutionizing the world of chess buffs. The changes are noticeable on diverse fronts. Whereas only ten years ago it was necessary to go to the local club to play a game of chess, now it is possible to do so from any spot one likes. This leads enthusiasts to play more, experiment more with favorite openings, and accelerate their learning curves.

Most games played on these servers are very fast. They are so fast that five-minute games are considered almost slow. Speed develops the tactical vi-

THOUSANDS OF INTERNET PLAYERS
VS. KASPAROV

Figure 5.20
Kasparov, playing white, versus the rest of the world. An experience only possible in the Internet era. Position after move 51 by white. Here, (51) . . . ♔a1! could have been played.

sion enormously, but it creates many problems related to addiction, since the game changes from an exercise of deep introspection and understanding to a game where the only objective is victory based on the fight against the clock. There is a category called *bullet games* that are played at a rhythm of one minute per player. The adrenalin rush these games provide leads to addiction among weak and not-so-weak players, who can spend hours at a time playing hundreds of these mostly senseless games. This is related to the self-centered charge that chess provides. For those who are living them at the moment, winning one game after another (even though they are technically disastrous and full of errors) continues to mean the triumph of one mind over another and the trampling of one will on the proposals of the other. It is an addiction that shows how absurd the human condition can be.

But besides speed chess, the Internet allows an infinity of activities. While I correct the galley proofs of the Spanish version of this book, I am watching live, from my own house, the broadcast of the 2005 World Chess Championship that is being played in San Luis, Argentina, thousands of miles away. My connection to the International Chess Club allows me to see all the games in real time, to have a discussion with thousands of chess enthusiasts, and to follow the commentaries of masters of different nationalities. With one click, I can enter databases with millions of games and read about the latest novelties in opening theory. If I feel like it, I could take a chess class with a grand master without leaving the room. Thanks to the Internet, the panorama of chess has changed much in the last few years and no doubt will change in ways that we can hardly imagine.

Summary

It is a universal human desire to understand how we learn to relate to the world that surrounds us, how our thoughts work, and so on in an endless line of questions, and to do so, we have generated metaphors of games, of perceptive processes, and of memory functioning. The founders of artificial intelligence believed in the computability of the intellect and learning and used chess as a testing ground for modeling the mind. But little by little, the minimax search algorithm, the most powerful chess programming tool, gained ground, a tribute to the immensity of the search space of moves in chess. Since then, chess programming has been seeking strategies that make minimax a more and more effective algorithm. The alpha-beta pruning technique, the use of hash tables, and the reordering of search tree variables are some of the elements that, along with representation techniques and the evaluation function, constitute the three main axes around which chess programs are formed. The machine has finally triumphed over human chess players, while the Internet has provided us with new ways of relating to each other in the chess community.

With these metaphors, we have almost reached the end of our trip. All that is left is to look at the final balance and ask ourselves again about the meaning of chess as a metametaphor.

Epilogue

For me, the passed pawn possesses a soul, just like a human being; it has unrecognized desires which slumber deep inside it and it has fears, the very existence of which it can but scarcely divine.

—*Aron Nimzowitsch*

The Creative Species, Player of Games

How has chess been able to touch so many vertices, so many metaphors of the human essence? The known and the strange, the feared and the adored, the respected and the disdained—all begin and end with how we represent the world and how we use that vision. The complexities of the universe are reflected in the complexities of our brains and in that natural, intimate and solitary activity that we call *mind*. In this process of matching up and representing, the inexhaustable human curiosity accepts the ancestral challenge of exploring the enormity of what we have yet to know. Chess, a world of fixed rules but with almost infinite borders, is an approachable model of that profound and endless human search.

The child who speaks to his teddy bear is learning to grant to those separate from himself the possession of thinking minds with their own desires and wills. For him, the teddy bear's mind is as real as his own. This practice will allow him to relate to other children in the game of life. The metaphor of one mind encountering another, a metametaphor that goes beyond games, suggests the idea of an openly curious humanity—of an animal that plays alone, with others, and with the environment to understand how things happen, what they are made of, and how they can be obtained, modified, or destroyed. In the process, human beings try to understand the environment, their own nature, and that of fellow humans.

The history that enriches the chess board transports us to ancient China and the divinatory arts, to ancient India and an army of chariots and elephants, to the Café de la Régence of eighteenth-century Paris and the creation of the modern state, and to early twentieth-century Vienna where the elite of science, culture, and politics can be found perusing the board and moving pieces.

Like chess players facing uncountable boards, human societies for millennia have been confronted with the game of life, learning to love, hate, and survive and trying to lose their fear in this strange world full of uncertainties. In every human act, there is an attempt to communicate a state of personal consciousness and to project it onto the conscious states of the rest of the world. Everything passes through the sieve of expectations—those that one has for oneself and for others and those that one believes others have for oneself. With (1) e4 e5, the game begins, and the proposals from each player's mind will become a conflict of wills.

Chess takes us—it *moves* us—from one extreme to the other of the endless spectrum of the human condition. During the "battle of ideas," as chess master Anthony Saidy would say, all kinds of metaphors, parables, and paradoxes occur. The game ignites brain signals that transmit a multitude of sensations. The star protagonist, our mind, time and again experiences the transcendence of a developed plan and the triviality of a forced move. We are gods, we are heroes, we are champions in every move where we can sense our victory, and in the next instant we fall into the deepest abyss of imminent loss. Throughout the length and breadth of the exploratory path, our mind passes from a state of passionate exaltation to profound depression. That is chess. That is life.

As social and cultural animals, we have received voluminous loads of popular wisdom, dogmas, beliefs, taboos, and knowledge. We have learned to recognize fear, to be suspicious and mistrustful, and to stay calm or become worried when faced with what is foreign to us. We have heard our grandparents' sayings and graduated from the family school, which made us experts in negotiating the possession of a toy with our siblings and bending our parents' will with a smile or tear. Our classmates, our teachers, our circle of friends, and our partners have molded our spirit and our way of being, our desires and our tastes. And our children remind us where we came from. They make us relive a stage of our lives that we cannot manage to call up exactly in our memory but that we know for certain is where we learned the rudiments of this social game that we move in every day.

In chess, our perception of the other is patent. In every calculation of every variation, our thoughts are a projection of what our opponent is thinking: "I know that he knows that I know." Two minds, isolated together, analyze and propose different answers to the problems that emerge from each position. For each position, there are two different visions. Alternative views cross over the same landscape. To the complexity of the position, we add the complexity of our own thoughts and then add the complexity of our opponent's thoughts. Fear of our rival's move (the threat) feeds ghosts that conjure up possibilities that are present in the choice of each move. What could be more human than a state of permanent doubt when facing the thoughts and actions of our fellow beings? With each move, chess reminds us that we live surrounded by a social fabric whose subtleties we have no choice but to understand and, as far as possible, to approximate with thought.

Machine versus Mind, Mind versus Machine

How can the enormous depth of a human being's social consciousness be encapsulated in a mathematical algorithm, a few production rules, or a set of recursive subroutines? Our trip through the metaphors that accompany chess has allowed us to visit the brain, the mind, and the cognitive processes. We have seen the digital revolution and the promises of artificial intelligence, some cognitive details of the player's mind, and the ways that a chess program is structured. What has been achieved after fifty years of applying basic science to chess? Chess programs have a brief history. The initial optimism about artificial intelligence has changed to the economic pragmatism of current chess programs, which rest on massive search algorithms and databases of openings and endgames.

We need to ask ourselves what we require to reproduce the human essence. Our brain holds a representation—a map—of our body. In the somatosensory cortex, all parts of our being—our hands, our arms, our organs—are connected directly or indirectly by dendrites and axons that create a network of synaptic connections, helping to form images of our life, our past, of present sensations and to create a representation of the possible futures that we might face. Those sensations come from our body to form our proprioception, a measure of our individuality in the face of the external environment (the beings and things that share our life).

We have seen that machines are now capable of playing like grand masters. They have a quantity of knowledge (in the form of evaluation functions, databases, and bitboards) that is much larger than human beings will ever be able to accumulate. Their capacity for calculation and memory cannot be equaled by a living organism. But their circuits hold no historical memory. Their processors leave the factory as functional units from an assembly line. The processor does not start its existence as a simple entity that acquires complexity during development. It does not grow or develop. And lacking development, it lacks the opportunity either to create its own representation for itself or to represent the parts of the computer to which it is connected.

For consciousness to emerge from an artificial system, the system will need to grow, and its most intimate components will need to undergo a process of self-organization. When this happens, the necessary conditions will exist for electrons' capacity to calculate in silicon to be translated into something infinitely more extraordinary—a capacity to represent itself and know that it exists as a singular entity separated from others.

Richard Reti or perhaps José Raúl Capablanca already said it. It is necessary to calculate only one move—the best—out of the twenty or thirty candidates for each position. The other moves are superfluous. And if knowledge (based on a grand master's experience of thousands of games) allows this miracle to happen (that the great, unique move stands out from the twenty or thirty candidates for each position), it is because the mind has elaborated a heuristic, a decision-making system of huge magnitude that bears witness to evolution's greatest achievement on earth—human intelligence.

How Does a Grand Master Think?

What is it like to be a bat? Philosopher Thomas Nagel's question is difficult to answer and perhaps even to ask. Consciousness has a personal character and is intimately linked to our unconscious proprioception. Our sense of individuality goes beyond our thoughts. It is recorded in every inch of our body and forms a totality that we cannot renounce. But how would a machine think? How would its circuits, input and output units, operating system, peripherals, and memory feel? How, if it could, would it represent all this in a series of 0s and 1s that are spread throughout its integrated silicon circuits? Would it play chess in the same way that it does now when automatism is the only process that exists within its processor?

If philosophy of the mind can ask what the existential experience of being a bat feels like, can we ask ourselves how a grand master thinks? Clearly we can, but we must admit that we will never be able to enter the mind of Garry Kasparov, share the thoughts of Judit Polgar, or know what Max Euwe thought when he discussed his protocol with Adriaan de Groot. If we really want to know how a grand master thinks, it is not enough to read Alexander Kotov, Nikolai Krogius, or even de Groot himself.

This book's explorations of the brain mechanisms and cognitive processes that accompany the act of decision making have opened the door to chess players' imagery. The activation of the different areas of the brain during a chess game opens a new dimension where thought and physiological activity can be followed at the same time. But if we really want to know how a grand master thinks, there is only one sure path: put in the long hours of study that it takes to become one. It is easier than trying to become a bat.

Emergent Processes

The emergence of conscious thought happened after millions of years of animal evolutionary history. More than 300 million years have passed since vertebrates began to leave the oceans and populate terrestrial ecosystems. Before birth, each human being recapitulates during embryonic development in the womb certain stages of the evolutionary phases of our ancestors (fish, amphibians, reptiles, and mammals). Our brains—with their lobes and circumvolutions, their millions of neurons, and their exquisite cocktails of neurotransmitters that travel across the synaptic buttons—are reconstructions of those that our ancestors had. We all descend from the brains of the African Eve and also the shrew (which, by escaping the claws of dinosaurs, allowed the great evolutionary expansion of the mammals), and their brains are present in ours. Even the molecules that help the embryo of our species to develop descend from molecules that are already present in groups as remote as flies and worms. Consciousness—that emergent process of the functioning of the human brain—is fueled by this ancestry as much as by the individual, personal development that we each undergo from birth until death.

In emergent processes, the whole is greater than the sum of the parts. A mathematical phenomenon that appears in certain dynamic systems also occurs within biological systems, from molecular interactions within the cells to the cognitive processes that we use to move within society. The emergence of consciousness from physical and chemical events that take place in our nervous system appears thanks to a slow process of development that begins with unconscious proprioception and culminates with an understanding of individuality. Our trip has reached its end. We are left with one message to convey: emergent processes also exist in chess. Emergent patterns of ideas, beauty, desires, or tragicomedy wait, ready to trap the next traveler in their complex domain of neatly patterned squares—the never-ending world of chess metaphors.

Appendixes

A   The Rudiments of Chess

Although this book uses chess as a lens for examining ideas about the brain, cognitive processes, and artificial intelligence, many readers might not be familiar with certain details of the game. This appendix covers some concepts that can help readers to take full advantage of the book's contents. The following section assumes that the reader knows at least the basic rules of chess: the goal of the game is to checkmate the opposing king, and certain rules control how pieces move, how pieces are exchanged, and how pawns are promoted. If not, the reader can visit the Internet addresses noted at the end of this appendix or consult hundreds of books for the beginning player. *Chess Fundamentals* by the Cuban world champion José Raúl Capablanca continues to be a superb introduction to the extraordinary world of chess. For the English-speaking world, there are a plethora of books to choose from, including Fred Reinfeld's or Bruce Pandolfini's many writings. In addition, a good place to start is the chess article in *Wikipedia*.

Notation

The system of chess notation has evolved from a descriptive one (noting a move, for example, as "queen's knight takes king's pawn," abbreviated QN x KP) to an algebraic system based on Cartesian coordinates that locate the pieces on the board. The files (columns) are identified by the letters *a* to *h*, and the ranks (rows) are numbered from 1 to 8 (figure A.1). In modern notation, the same move would be written Nxe5. Sometimes, as in this book, the symbol *N* is replaced by a drawing of the piece (♘xe5). This avoids confusion in translations. In English, for example, the knight is N (since the K is used for the king), while in Spanish the knight is C (for *caballo*). To avoid these problems, it was decided to replace the abbreviations with figures.

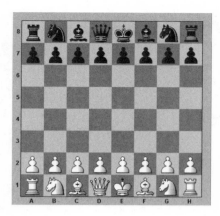

Figure A.1
A chess board coordinates system.

Other symbols that are used to comment games:

x   capture

0–0   kingside castling

0–0–0   queenside castling

+   check

++, #   checkmate

=   equal position

+=   white is slightly better

=+   black is slightly better

±   white is much better

!   good move

?   bad move

!?   interesting move

?!   dubious move

Terms

Center and Wing

The *center* is formed by the four squares where the main diagonals cross—e4, e5, d4, and d5. It is fundamental to dominate these squares to gain space, since a piece is able to control a large number of squares from the center, and to exert pressure on the opposite side. On both sides of the center along the files are the board's *wings* (also called *flanks*). The a, b, and c files are the queenside wing, and the f, g, and h files are the kingside. Note that visually white has the queenside wing to its left and the kingside wing to its right whereas black has the kingside to its left and the queenside to its right.

Combination: Forced and Winning

A *combination* is a sequence of moves that leads to the gain of some element, such as space (squares) or material (pawns and pieces) through a surprise move (sometimes a sacrifice of a pawn or piece). A *forced combination* is a series of moves that the opponent must follow or fall into a losing position. A *winning combination* determines the fight right away, either due to a checkmate or a major gain of material.

Elo Rating System

In chess's competitive facet, each player's strength can be quantitatively evaluated, with the Elo rating system, which is based on the work of American physicist Arpad Elo. Elo scoring objectively quantifies players' relative strengths. The standard deviation of 200 points is the approximate separation between classes of players. The U.S. Chess Federation has four major classes—class A players (who have a rating from 1800 to 2000), expert players (2000 to 2200), masters (2200 to 2400), and senior masters (2,400 and above). The World Chess Federation (FIDE, for Fédération Intérnationale des Echecs) awards the international master (IM) title for ratings of 2400 up to 2500 and the grand master (GM) title for 2500 and above (after achieving the appropriate norms). The unofficial super grand master category is reserved for players who have broken the barrier of 2700 points. At present, twenty players are on this list.

En Prise

A piece that is *en prise* is said to be exposed to imminent capture.

Fianchetto and Fianchettoed Bishop

Fianchetto is a characteristic move of the knight's pawn. After advancing a square, it leaves an open space for the bishop (the *fianchettoed bishop*) so that it can be developed on the long diagonal. This kind of formation is appreciated by the attacking side as well as the defender and constitutes the basis of many openings and defenses (such as the Reti opening and the Nimzo-Indian defense).

Gambit

A *gambit* is a temporary sacrifice of a pawn in the initial phases of the opening to gain an advantage in the development. The queen's gambit (1) d4 d5, (2)c4 . . . is one of the classic openings in master-level chess. Others, like the king's gambit (1) e4 e5, (2)f4, have fallen into disuse but were popular in chess's romantic era since they often lead to an open and spectacular game.

Line and Main Variation

A *line* is a sequence of possible moves that can be developed from a given position. The *main line* or *variation* is the one that a player would most typically use.

Maneuver

A *maneuver* is a sequence of moves that leads to the pieces being better positioned.

Motif, Theme

A *motif* or *theme* is a standard position of the pieces from which they can carry out typical maneuvers or combinations. Classic motifs include a minority attack on the queenside, the weakness of the f7 pawn, a mobile center, an isolated queen's pawn, and an open c file.

Opening, Middle Game, and Endgame

The *opening, middle game,* and *endgame* are the three main stages that divide the game. As an interesting side note, there is a correspondence between the three phases of chess and the three acts of classical theater (exposition, confrontation, and resolution). In the *opening*, the player tries to develop the pieces, control the center, and move the king to safety by castling.

An immense body of opening theory exists. An interesting and as yet unresolved debate concerns how much advantage white gains from the fact of moving first. Normally, black plays to neutralize this small advantage, while white tries to capitalize on it. The list below provides a basic guide to openings with their corresponding ECO (Encyclopaedia of Chess Openings) code. Openings are divided into five groups—A, B, C, D, and E. Each group has ninety-nine subgroups, and in each subgroup there are different variations. For example, the Sicilian defense (1) e4 c5, one of the most popular systems for black, consists of eighty variations from B20 to B99.

A00	Irregular openings
A01	Nimzovitch-Larsen attack (1) b3

A02–A03   Bird's opening (1) f4

A04–A09   Reti opening (1) ♘f3

A10–A39   English opening (1) c4

A40–A41   Queen's pawn game (1) d4

A42   Averbakh system (1) d4 d6, (2) c4 g6, (3) ♘c3 ♗g7, (4) e4

A43–A44   Old Benoni defense (1) d4 c5

A45–A46   Queen's pawn game (1) d4 ♘f6

A47   Queen's Indian defense (1) d4 ♘f6, (2) ♘f3 b6

A48–A49   King's Indian defense (1) d4 ♘f6, (2) ♘f3 g6

A50   Queen's pawn game (1) d4 ♘f6, (2) c4

A51–A52   Budapest defense (1) d4 ♘f6, (2) c4 e5

A53–A55   Old Indian defense (1) d4 ♘f6, (2) c4 d6

A56–A79   Benoni defense (1) d4 ♘f6, (2) c4 c5

A80–A99   Dutch defense (1) d4 f5

B00   King's pawn opening (1) e4

B01   Scandinavian (center counter) defense (1) e4 d5

B02–B05   Alekhine defense (1) e4 ♘f6

B06   Robatsch defense (1) e4 g6

B07–B09   Pirc defense (1) e4 d6, (2) d4 ♘f6, (3) ♘c3

B10–B19   Caro-Kann defense (1) e4 c6

B20–B99   Sicilian defense (1) e4 c5

C00–C19   French defense (1) e4 e6

C20   King's pawn game (1) e4 e5

C21–C22   Danish gambit (1) e4 e5, (2) d4 exd4

C23–C24   Bishop's opening (1) e4 e5, (2) ♗c4

C25–C29   Vienna game (1) e4 e5, (2) ♘c3

C30–C39   King's gambit (1) e4 e5, (2) f4

C40   King's knight games (1) e4 e5, (2) ♘f3

C41   Philidor's defense (1) e4 e5, (2) ♘f3 d6

C42–C43   Petroff's defense (1) e4 e5, (2) ♘f3 ♘f6

C44   King's pawn game (1) e4 e5, (2) ♘f3 ♘c6

C45   Scotch game (1) e4 e5, (2) ♘f3 ♘c6, (3) d4 exd4, (4) ♘xd4

C46   Three knights game (1) e4 e5, (2) ♘f3 ♘c6, (3) ♘c3

C47–C49   Four knights game (1) e4 e5, (2) ♘f3 ♘c6, (3) ♘c3 ♘f6

C50   King's pawn game (1) e4 e5, (2) ♘f3 ♘c6, (3) ♗c4

C51–C52   Evans gambit (1) e4 e5, (2) ♘f3 ♘c6, (3) ♗c4, (4) c4

C53–C54   Giuoco piano (1) e4 e5, (2) ♘f3 ♘c6, (3) ♗c4 ♗c5, (4) c3

C55–C59   Two knights defense (1) e4 e5, (2) ♘f3 ♘c6, (3) ♗c4 ♘f6

C60–C99   Ruy López (Spanish opening) (1) e4 e5, (2) ♘f3 ♘c6,
(3) ♗b5

D00–D05   Queen's pawn game (1) d4 d5

D06–D69   Queen's gambit (1) d4 d5, (2) c4

D70–D79   Neo-Gruenfeld defense (1) d4 ♘f6, (2) c4 g6, (3) f3 d5

D80–D99   Gruenfeld defense (1) d4 ♘f6, (2) c4 g6, (3) ♘c3 d5

E00   Queen's pawn game (1) d4 ♘f6, (2) c4 e6

E01–E09   Catalan opening (1) d4 ♘f6, (2) c4 e6, (3) g3 d5

E10   Queen's pawn game (1) d4 ♘f6, (2) c4 e6, (3) ♘f3

E11   Bogo-Indian defense (1) d4 ♘f6, (2) c4 e6, (3) ♘f3 ♗b4+

E12–E19   Queen's Indian defense (1) d4 ♘f6, (2) c4 e6, (3) ♘f3 b6

E20–E59   Nimzo-Indian defense (1) d4 ♘f6, (2) c4 e6, (3) ♘c3 ♗b4

E60–E99   King's Indian defense (1) d4 ♘f6, (2) c4 g6

The *middle game* follows the opening. It is characterized by a fight among the pieces, which are now in different squares from where they started. Middle-game maneuvers are full of subtleties. Depending on the type of opening that was played, positions generally are reached where there are open diagonals and files (with a tactical game) or positions where everything is more closed (with a more strategic game).

When only pawns and a few pieces are left on the board, the game enters the *endgame*, the final phase. This stage is characterized by the possibility of promoting pawns in the eighth rank, and thus the fight centers on advancing one's own pawns and avoiding the advance of the opponent's pawns.

Each phase of the game has a series of different values. In the endgame, for example, pawns can become more valuable than pieces, while in the middle game the loss of a piece generally results in the loss of the game.

Pawn Chain

A *pawn chain* is a group of contiguous, same-color pawns that defend or can defend each other (figure A.2).

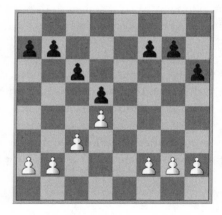

Figure A.2
Pawn chains. In this example, each side has two pawn chains.

Pawn: Passed, Isolated, and Backward

A *passed pawn* has advanced in front of the opposing pawns and can be stopped only by opposite pieces. A classic maneuver to prevent its advance is to block the passed pawn with a minor piece, usually a knight. Passed pawns are dangerous because they constantly threaten to be promoted in the eighth rank. An *isolated pawn* cannot be protected by other pawns and thus usually constitutes a weakness in the final phase of the game. A *backward pawn* has lagged behind its accompanying pawns and cannot be defended by them. It is a special case of an isolated pawn.

Position

A *position* is the situation of the pieces on the board at a given point in the game. The position determines the type of strategy that is carried out. A *closed position* has no open files or diagonals and requires the game to be played positionally, whereas an *open position* has diagonals and files opened leading to tactical maneuvers.

Ranks, Files, and Diagonals

Ranks are the horizontal lines of squares and are numbered from 1 to 8. Expressions such as "Seize the seventh and eighth ranks" are classic strategic plans for the middle game and endgame. *Files* are the vertical lines of squares and are identified by letters from *a* to *h*. An open file does not have pawns and allows the major pieces (rooks and queen) to be developed without any obstacle. A half-open file has only one pawn. *Diagonals* are squares of the same color that run obliquely across the board; they are bishops' natural paths.

Sacrifice

A *sacrifice* is a play that gives up material (a pawn or piece) to secure a determined objective, either regaining material with an improved position or leading directly to a winning position. Sacrifices have an important aesthetic appeal due to the emotional weight that accompanies them. Players who carry out a sacrifice are committed to the line that they have chosen, and any mistake in calculation would be fatal.

Simultaneous, Simul

A *simultaneous* or *simul* is a series of exhibition games played by one strong player at the same time. The player moves from board to board making moves. The player sometimes plays with his back to the boards (called *blindfold playing*), keeping information about each of the games he is playing in his memory. In a regular simul, a player can top 100 games, and in a blindfold simultaneous, players have reached more than fifty.

Strategy, Ideas, and Plans

Strategy, *ideas*, and *plans* are concepts that guide a player's objectives and that are determined by the characteristics of the position. They are the heuristics of expert players. An *idea* quickly suggests a series of possible moves and leads to others being discarded.

Tactics

Tactics are made up of maneuvers and combinations. They are the means of carrying out a strategic plan.

Tempo

A *tempo* is equivalent to one player's move. It is called a *ply* in chess programming. This book uses the term *move* both for one player's turn and also for both black and white's turns, depending on context. In endgames, it is important to count tempos exactly, since the game is decided by who can promote a pawn first.

Weak Squares and Holes

Weak squares cannot be defended by their own pawns because they have advanced and created *holes* in which the opponent can place pieces without fear of being expelled by the opposing pawns.

Zugzwang

Zugzwang is the German term for a position where the side whose turn it is loses no matter what move is played. This type of situation happens in endgames when few pieces remain on the board. There are several techniques for trying to force an opponent into *zugzwang*, such as maneuvering backward with the king to lose a tempo.

B   Chess Programs and Other Tools

Artificial intelligence has led to numerous expert systems that are used in a variety of disciplines (including clinical diagnoses, elaboration of molecular substances, and fire prevention). Nowadays, chess programs play at levels far beyond an average player's abilities. These programs have enormous speed, the capacity to store millions of positions, increasingly effective search algorithms and evaluation functions, and enormous databases. All this has lifted chess programs to the level of grand masters. For the past ten years, the Swedish Association of Chess Programs has maintained a list of the relative strengths of more than 250 commercial programs, which they have play each other under tournament conditions and participate in tournaments where they play human competitors. The list of the top fifty programs for 2005 was impressive. Two had passed the barrier of 2800 Elo rating points, something that as of 2008 had been matched only by Garry Kasparov, Viswanathan Anand, Vladimir Kramnik, and Veselin Topalov. In the group's November 2007 list, two versions of Rybka, a new kid on the block, were rated over 2900, while twelve other programs were above 2800. Two sites offer free programs for downloading—the powerful Crafty (at ftp://ftp.cis.uab.edu/pub/hyatt/ and also at www.craftychess.org) and the GNU program (at http://www.gnu .org). In addition, read the *Wikipedia* article on chess engines for its information and links, and visit Tim Mann's site at www.tim-mann.org; they are both great places to start looking.

Besides chess programs, computer science has given another great tool to the world of chess—chess databases that archive millions of games. These databases make it possible to analyze the games of current masters without spending all the time previously necessary to track them down. Most of the games of international matches are published on the Internet almost as they are being played. New theories of openings travel now at lightning speed through cyberspace.

ChessBase has become a classic database program. Its impressive functionality lets users look for every occurrence of a given position throughout all the games in the database or create a number of filters to create, organize, and rearrange subbases of certain topics. In addition, thanks to the option to load chess programs (like Fritz), ChessBase allows games to be analyzed and have new variations played with them.

The most frequently used format for exchanging game files is called PGN (portable game notation). Thanks to this simple, universal format, games can be transferred between different chess programs or databases. A heading presents commentaries and information about the game, and the movements are written in algebraic notation. Since PGN is in uncompressed ASCII format, it occupies a lot of space for a database with hundreds of thousands of games. On the other hand, it can be created and read with any text editor. Many free programs and tools are available online for creating, publishing, visualizing, and analyzing databases in PGN. The best-known is Winboard (http://www.tim-mann.org/chess.html), a versatile graphical interface that can read games from PGN files and can load chess playing programs (as ChessBase does).

C Internet Sites for Playing Chess

Hundreds and sometimes thousands of people throughout the world can be found on Internet sites playing chess at any hour of the day or night. The only requirement is a computer with an Internet connection and the ability to understand and tap out a little English, since most sites are conducted in that language. Many sites are currently available for playing—from general servers (like Yahoo! or Playsite) to sites that are dedicated exclusively to chess. Among the latter, Free Internet Chess Server (FICS), Internet Chess Club (ICC), and the Spanish-language server Ajedrez21 stand out:

- Free Internet Chess Server (FICS) (www.freechess.com) is a free, English-language site for chess playing.
- Internet Chess Club (ICC) (www.chessclub.com) is the site par excellence for chess playing. It allows players to compete against players worldwide and in any Elo category. The club also offers skills classes (lessons can be viewed in flash format at www.chess.fm) and commentary on chess news and events. Many grand masters play on this site, and users can watch grand masters' games live. The server offers several channels in different languages, including Spanish.
- Ajedrez21 (www.ajedrez21.com) was a Spanish chess-playing site that was run by Spanish grand master Miguel Illescas's Academy. A21 has integrated as a Spanish channel at ICC. A new free site called Buho21 (www.buho21.com) has been developed by the same programmers and is currently active, although it is not related to Illescas's school.

In addition to the playing sites, which incorporate all kinds of information about chess, the Internet contains repositories of chess theory, chess programming, and chess games. Two of the top sites are Chess Lab and ChessBase:

- Chess Lab (www.chesslab.com) is an Internet site where the user can introduce a position and access a database of more than 2 million games that are updated weekly. The database pulls up all games in which a similar position has been reached. Users can access a game room and analyze their own positions and play against a program.
- ChessBase (www.chessbase.com) is the site of the most important commercial programs on the market, including ChessBase's own database and programs like Fritz and Junior. It has a database that can be used online. It is also an active playing site.

Bibliography

This bibliography lists some of the many books and specialized magazines that I used to check information. Interested readers should find this bibliography a useful starting point for moving more deeply into the proposed metaphors. As if I were moving through a chess search tree, reading one article opened up dozens of new questions that led to reading another dozen articles and so on nearly ad infinitum, like Borges's library of Babel. This exponential explosion of material means that it is impossible for me to mention more than a portion of the sources consulted. I owe a great debt to these authors. I have commented briefly on some that seem most important from the perspective of this book. Some of these references were read in Spanish, and that version is listed. Finally, although I have tried to divide the references by chapter, some overlapping is inevitable.

Chapter 1

Arsuaga J. Ll, and I. Martínez. *La especie elegida: La larga marcha de la evolución humana*. Madrid: Temas de Hoy, 1998. Analysis of the most important hominid fossil discoveries and human evolution, from two of the Atapuerca site researchers.

Atherton, M., J. Zhuang, W. M. Bart, X. Hu, and S. He. "A Functional MRI Study of High-Level Cognition. I. The Game of Chess." *Brain Research and Cognitive Brain Research* 16, no. 1 (2003): 26–31.

Callebaut, W., and D. Rasskin-Gutman, eds. *Modularity: Understanding the Development and Evolution of Natural Complex Systems*. Cambridge: MIT Press, 2005. The concept of modularity in the arts and sciences, with a posthumous preface by Herbert Simon and a chapter by Fernand Gobet dedicated to chess.

Chabris, C. F., and S. E. Hamilton. "Hemispheric Specialization for Skilled Perceptual Organization by Chessmasters." *Neuropsychologia* 30, no. 1 (1992): 47–57.

Chen, X., D. Zhang, X. Zhang, Z. Li, X. Meng, S. He, and X. Hu. "A Functional MRI Study of High-Level Cognition. II. The Game of GO." *Brain Research and Cognitive Brain Research* 16, no. 1 (2003): 32–37.

Dalgleish, T. "The Emotional Brain." *National Review of Neuroscience* 5, no. 7 (2004): 583–589.

Edelman, G. E.. *Bright Air, Brilliant Fire*. New York: Penguin, 1992. Fascinating analysis of the central nervous system from the point of view of the dynamics of development and evolutionary processes.

Glimcher, P. W., and A. Rustichini. "Neuroeconomics: The Consilience of Brain and Decision." *Science* 306, no. 5695 (2004): 447–452.

Houde, O., and N. Tzourio-Mazoyer. "Neural Foundations of Logical and Mathematical Cognition." *National Review of Neuroscience* 4, no. 6 (2003): 507–514.

Kaas, J. H., and C. E. Collins. "Evolving Ideas of Brain Evolution." *Nature* 411 (2001): 141–142.

Lamprecht, R., and J. LeDoux. "Structural Plasticity and Memory." *National Review of Neuroscience* 5, no. 1 (2004): 45–54.

Langton, C. G., ed. *Artificial Life: An Overview*. Cambridge: MIT Press, 1997.

Mann, M. W., B. Gueguen, S. Guillou, E. Debrand, and C. Soufflet. "Chess-Playing Epilepsy: A Case Report with Video-EEG and Back Averaging." *Epileptic Disorders* 6, no. 4 (2004): 293–296.

Nichelli, P., J. Grafman, P. Pietrini, D. Alway, J. C. Carton, and R. Miletich. "Brain Activity in Chess Playing." *Nature* 369, no. 6477 (1994): 191.

Onofrj, M., L. Curatola, G. Valentini, M. Antonelli, A. Thomas, and T. Fulgente. "Nondominant Dorsal-Prefrontal Activation during Chess Problem Solution Evidenced by Single Photon Emission Computerized Tomography (SPECT)." *Neuroscience Letters* 198, no. 3 (1995): 169–172.

Passingham, R. E., K. E. Stephan, and R. Kotter. "The Anatomical Basis of Functional Localization in the Cortex." *National Review of Neuroscience* 3, no. 8 (2002): 606–616.

Saxe, R., S. Carey, and N. Kanwisher. "Understanding Other Minds: Linking Developmental Psychology and Functional Neuroimaging." *Annual Review of Psychology* 55 (2004): 87–124.

Serruya, M. D., G. Nicholas, N. G. Hatsopoulos, L. Paninski, M. R. Fellows, and J. P. Donoghue. "Instant Neural Control of a Movement Signal." *Nature* 416 (2002): 141–142.

Simon, H. A. "The Architecture of Complexity." *Proceedings of the American Philosophical Society* 106 (1962): 467–482.

Toga, A. W., and P. M. Thompson. "Mapping Brain Asymmetry." *National Review of Neuroscience* 4, no. 1 (2003): 37–48.

Tononi, G., G. M. Edelman, and O. Sporns. "Complexity and Coherency: Integrating Information in the Brain." *Trends in Cognitive Sciences* 2 (1998): 474–484.

Chapter 2

Albright, T. D., E. R. Kandel, and M. I. Posner. "Cognitive Neuroscience." *Current Opinion in Neurobiology* 10, no. 5 (2000): 612–624.

Andrews, P. W. "The Psychology of Social Chess and the Evolution of Attribution Mechanisms: Explaining the Fundamental Attribution Error." *Evolution of Human Behavior* 22(1) (2001): 11–29.

Baddeley, A. "Working Memory." *Science* 255, no. 5044 (1992): 556–559.

Baddeley, A. "Working Memory: Looking Back and Looking Forward." *National Revview of Neuroscience* 4, no. 10 (2003): 829–839.

Cacioppo, J. T. "Common Sense, Intuition, and Theory in Personality and Social Psychology." *Personal Sociology Psychological Review* 8, no. 2 (2004): 114–122.

Cairns-Smith, A. G. *Evolving the Mind: On the Nature of Matter and the Origin of Consciousness*. Cambridge: Cambridge University Press, 1998.

Damasio, A. *The Feeling of What Happens: Body and Emotion in the Making of Consciousness*. New York: Harcourt Brace, 1999. Important contribution from the field of neuromedicine to the meaning of consciousness and its relation to emotions.

Damasio, A. R. "How the Brain Creates the Mind." *Scientific American* 281, no. 6 (1999): 112–127.

Deary, I. J. "Human Intelligence Differences: A Recent History." *Trends in Cognitive Science* 5, no. 3 (2001): 127–130.

Deary, I. J. "Human Intelligence Differences: Towards a Combined Experimental-Differential Approach." *Trends in Cognitive Science* 5, no. 4 (2001): 164–170.

Deary, I. J., and P. G. Caryl. "Neuroscience and Human Intelligence Differences." *Trends in Neuroscience* 20, no. 8 (1997): 365–371.

Dehaene, S., M. Kerszberg, J. P. Changeux. "A Neuronal Model of a Global Workspace in Effortful Cognitive Tasks." *Proceedings of the National Academy of Sciences USA* 95, no. 24 (1998): 14529–14534.

Ericsson, K. A., and W. Kintsch "Long-Term Working Memory." *Psychology Review* 102, no. 2 (1995): 211–245.

Gellatly, A., and O. Zarate. *Introducing Mind and Brain*. New York: Totem Books. Basic, 1999. A book on the structure and function of the brain written in cartoon style.

Gleitman, H. *Psychology*. New York: Norton, 1981. A psychology textbook with a good introduction to cognitive science.

Hobson, J. A. *Consciousness*. New York: Scientific American Library, 1999. An introduction to the concept of consciousness from the perspective of neuromedicine. Hobson is a specialist in the study of dreams.

Horn, G. "Pathways of the Past: The Imprint of Memory." *National Review of Neuroscience* 5, no. 2 (2004): 108–120.

Humphrey, N. *Consciousness Regained: Chapters in the Development of Mind*. Oxford: Oxford University Press, 1983.

Kandel, E. R. "The Molecular Biology of Memory Storage: A Dialogue between Genes and Synapses." *Science* 294, no. 5544 (2001): 1030–1038.

Kane, M. J. "The Intelligent Brain in Conflict." *Trends in Cognitive Science* 7, no. 9 (2003): 375–377.

Koch, C., and G. Laurent. "Complexity and the Nervous System." *Science* 284, no. 5411 (1999): 96–98.

Larkin, J. H., and H. A. Simon. "Why a Diagram Is Sometimes Worth Ten Thousand Words." *Cognitive Science* 11 (1987): 65–99.

Li, S. C., U. Lindenberger, B. Hommel, G. Aschersleben, W. Prinz, and P. B. Baltes. "Transformations in the Couplings among Intellectual Abilities and Constituent Cognitive Processes across the Life Span." *Psychological Science* 15, no. 3 (2004): 155–163.

Lisman, J. "Long-Term Potentiation: Outstanding Questions and Attempted Synthesis." *Philosophical Transactions of the Royal Society London B Biological Science* 358, no. 1432 (2003): 829–842.

McCaugh, J. L. "Memory: A Century of Consolidation." *Science* 287 (2000): 248–251.

Miller, G. A. "The Magical Number Seven, Plus or Minus Two: Some Limits on Our Capacity for Processing Information." *Psychological Review* 63 (1956): 81–97.

Minsky, M. *The Society of Mind*. New York: Simon & Schuster, 1986. An analysis of the mind as a modular complex, from one of the fathers of artificial intelligence.

Miyashita, Y. "Cognitive Memory: Cellular and Network Machineries and Their Top-Down Control." *Science* 306, no. 5695 (2004): 435–440.

Nakazawa, K., T. J. McHugh, M. A. Wilson, and S. Tonegawa. "NMDA Receptors, Place Cells and Hippocampal Spatial Memory." *National Review of Neuroscience* 5, no. 5, no. (2004): 361–372.

Newell, A., and H. A. Simon. *Human Problem Solving*. Englewood Cliffs, NJ: Prentice-Hall, 1972.

Penrose, R. *The Emperor's New Mind: Concerning Computers, Minds, and the Laws of Physics*. Oxford: Oxford University Press, 1989.

Peterson, L. R., and M. Peterson. "Short-Term Retention of Individual Items." *Journal of Experimental Psychology* 58 (1959): 193–198.

Plomin, R., and F. M. Spinath. "Genetics and General Cognitive Ability (g)." *Trends in Cognitive Science* 6, no. 4 (2002): 169–176.

Richman, H. B., and H. A. Simon. "Context Effects in Letter Perception: Comparison of Two Theories." *Psychological Review* 96 (1989): 417–432.

Richman, H. B., I. J. Straszewski, and H. A. Simon. "Simulation of Expert Memory with EPAM IV." *Psychological Review* 102 (1995): 305–330.

Robertson, L. C. "Binding, Spatial Attention and Perceptual Awareness." *National Review of Neuroscience* 4 (2003): 93–102.

Ruiz Vargas, J. M. *La memoria humana: Función y estructura*. Madrid: Alianza Editorial, 1994. Good, in-depth introduction to memory systems from a cognitive perspective.

Sapolsky, R. M. "Stressed-Out Memories." *Scientific American* (December 2004): 28–33.

Simon, H. A. "The Information-Processing Explanation of Gestalt Phenomena." *Computers in Human Behaviour* 2 (1986): 155–241.

Simons, J. S., and H. J. Spiers. "Prefrontal and Medial Temporal Lobe Interactions in Long-Term Memory." *National Review of Neuroscience* 8 (2003): 637–648.

Skoyles, J. R. "Evolution's 'Missing Link': A Hypothesis upon Neural Plasticity, Prefrontal Working Memory and the Origins of Modern Cognition." *Medical Hypotheses* 48, no. 6 (1997): 499–501.

Vogeley, K., P. Bussfeld, A. Newen, S. Herrmann, F. Happe, P. Falkai, W. Maier, N. J. Shah, G. R. Fink, and K. Zilles. "Mind Reading: Neural Mechanisms of Theory of Mind and Self-Perspective." *Neuroimage* 14, no. 1 (2001): 170–181.

Chapter 3

Adarraga, P., and J. L. Zaccagnini, eds. *Psicología e inteligencia artificial*. Madrid: Editorial Trotta, 1994. Good introduction to artificial intelligence concepts and applications from Spanish researchers who specialize in the generation of expert systems.

Arbib, M. A. *Cerebros, máquinas y matemáticas*. Madrid: Alianza Universidad, 1997. A classic introduction to cybernetics.

Boole, G. *El análisis matemático de la lógica*. Madrid: Cátedra, 1984. The book that brought logic and mathematics together.

Brooks, R. A. "Elephants Don't Play Chess: Robotics and Autonomous Systems" 6 (1990): 3–15.

Cuena, J., et al. *Inteligencia artificial: Sistemas expertos*. Madrid: Alianza Editorial, 1985. Articles about the classic expert systems.

Fischer, M. A., and O. Firschein. *Intelligence: The Eye, the Brain, and the Computer*. Reading, MA: Addison Wesley, 1987. A textbook that provides a knowledgeable (although somewhat out-of-date) look at the relation between human and artificial intelligence. Highly recommended as a general introductory text for understanding the basic problems in the field.

French, R. M. "The Turing Test: The First Fifty Years." *Trends in Cognitive Science* 4(3) (2000): 115–122.

Hartnell, T. *Inteligencia artificial: Conceptos y programas*. Madrid: Anaya Multimedia, 1984. Basic introduction to the concepts of artificial intelligence with concrete applications and programs written in BASIC.

Luck, M., P. McBurney, O. Shehory, and S. Willmott. *Agent Technology: Computing as Interaction. A Roadmap for Agent-Based Computing*. Southampton: University of Southampton, 2005.

Pfeifer, R., and J. C. Bongard. *How the Body Shapes the Way We Think: A New View of Intelligence*. Cambridge: MIT Press, 2006.

Rasskin, M. *Música Virtual*. Madrid: Anaya Multimedia, 1993. A lucid analysis of music making and computers.

Chapter 4

Aagaard, J. *Inside the Chess Mind: How Players of All Levels Think about the Game*. London: Everyman Chess, 2004. Interesting mix of interviews with professional and amateur players.

Amidzic, O., H. J. Riehle, T. Fehr, C. Wienbruch, and T. Elbert. "Pattern of Focal Gamma-Bursts in Chess Players." *Nature* 412, no. 6847 (2001): 603.

Archer, H. A., J. M. Schott, J. Barnes, N. C. Fox, J. L. Holton, T. Revesz, L. Cipolotti, and M. N. Rossor. "Knight's Move Thinking? Mild Cognitive Impairment in a Chess Player." *Neurocase* 11, no. 1 (2005): 26–31.

Avni, A. *The Grandmaster's Mind*. London: Gambit Publications, 2004. An excellent book that offers interviews with various players, some of them elite-level, narrating how they think in front of the chessboard. The author dissects the narratives with great skill and presents some interesting conclusions.

Avni, A., D. Kipper, and S. Fox. "Personality and Leisure Activities: An Illustration with Chess Players." *Personal and Individual Differences* 8, no. 5 (1987): 715–719.

Barry, H. "Longevity of Outstanding Chess Players." *Journal of Genetics and Psychology* 115 (1st half) (1969): 143–148.

Binet, A. "Mnemonic Virtuosity: A Study of Chess Players." *Genetics and Psychology Monographs* 74, no. 1 (1966): 127–162.

Binet, A. *Psychologie des Grand Calculateurs et des jouers d'échecs*. Paris: Hachette, 1894. Classic book with the first assessments of chess as a cognitive activity. It is available on the Internet at http://users.lk.net/~stepanov/mnemo/ binetf.html.

Bonsdorff, F., and R. Bonsdorff. *Ajedrez y matemáticas*. Barcelona: Ediciones Martínez Roca, 1976. Contains interesting facts about the geometry of the chessboard and the relation to the movement of pieces.

Charness, N. "Aging and Skilled Problem Solving." *Journal of Experimental Psychology and Genetics* 110, no. 1 (1981): 21–38.

Charness, N. "The Impact of Chess Research on Cognitive Science." *Psychological Research* 54 (1992): 4–9. A concise review that lays out the results and the scope of research in chess on cognitive science theories, especially memory, problem solving, and perception.

Charness, N. "The Role of Theories of Cognitive Aging: Comment on Salthouse." *Psychology of Aging* 3, no. 1 (1988):17–21.

Charness, N. "Visual Short-Term Memory and Aging in Chess Players." *Journal of Gerontology* 36, no. 5 (1981): 615–619.

Charness, N., E. M. Reingold, M. Pomplun, and D. M. Stampe. "The Perceptual Aspect of Skilled Performance in Chess: Evidence from Eye Movements." *Memory and Cognition* 29, no. 8 (2001): 1146–1152.

Chase, W. G., and H. A. Simon. "Perception in Chess." *Cognitive Psychology* 4 (1973): 55–81.

Cherington, M. "Visual Neglect in a Chess Player." *Journal of Nervous Mental Disorders* 159, no. 2 (1974): 145–147.

Cooke, N. J., R. S. Atlas, D. M. Lane, and R. C. Berger. "Role of High-Level Knowledge in Memory for Chess Positions." *American Journal of Psychology* 106 (1993): 321–351.

Cranberg, L. D., and M. L. Albert. "The Chess Mind." In L. Kober and D. Fein, eds., *The Exceptional Brain: Neuropsychology of Talent and Special Abilities* (156–190). New York: Guilford Press, 1988.

de Groot, A. *Thought and Choice in Chess*. The Hague: Mouton, 1965. A required reference for understanding the impact of chess on the cognitive sciences. This edition has transcriptions of the protocols of Alekhine, Euwe, Fine, Flohr, Keres, and Tartakower speaking out loud. An indispensable gem.

de Groot, A., and F. Gobet. *Perception and Memory in Chess: Heuristics of the Professional Eye*. Assen: Van Gorcum, 1996. A continuation of the classic written by de Groot. Brought up-to-date and including perception and visual attention experiments and Gobet's additional computational focus.

Djakow, I. N., N. W. Petrowski, and P. A. Rudik. *Psychologie des Schachspiels*. Berlin: de Gruyter, 1927.

Dunnington, A. *Chess Psychology: Approaching the Psychological Battle Both on and off the Board*. London: Everyman Chess, 2003. A personal vision of the factors that interfere with chess thinking.

Eales, R. *Chess: The History of a Game*. London: Batsford, 1985. A very good and highly erudite chess history.

Eisele, P. "Judgment and Decision-Making: Experts' and Novices' Evaluation of Chess Positions." *Perception and Motor Skills* 98, no. 1 (2004): 237–248.

Eisenstadt, M., and Y. Kareev. "Perception in Game Playing: Internal Representation and Scanning of Board Positions." In P. N. Johnson-Laird and P. C. Wason, eds., *Thinking: Readings in Cognitive Science*. Cambridge: Cambridge University Press, 1977.

Ericsson, K. A., and A. C. Lehmann. "Expert and Exceptional Performance: Evidence of Maximal Adaptation to Task Constraints." *Annual Review of Psychology* 47 (1996): 273–305.

Euwe, M. *Judgement and Planning in Chess*. New York: McKay Chess Library, 1979. An essential book that explains step-by-step the creation of chess plans.

Feigenbaum, E. A. "EPAM-like Models of Recognition and Learning." *Cognitive Science* 8 (1984): 305–336.

Fine, R. *Psicología del jugador de ajedrez*. Barcelona: Ediciones Martínez Roca, 1978. A psychoanalytical focus on chess. A controversial book that is relevant only in the light of Freudian theory and for the fact of having been written by one of the strongest grand masters of the mid-twentieth century.

Fine, R. "The Psychology of Blindfold Chess: An Introspective Account." *Acta Psychologica* 24 (1965): 352–370.

Finkenzeller, R., W. Ziehr, and E. M. Bührer. *Chess: A Celebration of 2000 Years*. London: Mackenzie, 1990. A standard encyclopedia that provides an acceptable introduction to chess.

Fischer, B. *My Sixty Memorable Games*. New York: Simon and Schuster, 1969.

Frey, P. W., and P. Adesman. "Recall Memory for Visually Presented Chess Positions." *Memory and Cognition* 4 (1976): 541–547.

Freyhoff, H., H. Gruber, and A. Ziegler. "Expertise and Hierarchical Knowledge Representation in Chess." *Psychological Research* 54 (1992); 32–37.

Frydman, M., and R. Lynn. "The General Intelligence and Spatial Abilities of Gifted Young Belgian Chess Players." *British Journal of Psychology* 83 (Pt. 2) (1992): 233–235.

Garland, D. J., and J. R. Barry. "Cognitive Advantage in Sport: The Nature of Perceptual Structures." *American Journal of Psychology* 104, no. 2 (1991): 211–228.

Gobet, F. "Can Deep Blue Make Us Happy? Reflection on Human and Artificial Expertise." *Papers from the 1997 AAAI Workshop. Deep Blue versus Kasparov: The Significance for Artificial Intelligence* (20–23). AAAI Press: Technical Report WS-97-04.

Gobet, F. "Chess Players Thinking Revisited." *Swiss Journal of Psychology* 57, no. 1 (1998): 18–32.

Gobet, F. "Expert Memory: A Comparison of Four Theories." *Cognition*. 66, no. 2 (1988): 115–152.

Gobet, F. "Learned Helplessness in Chess Players: The Importance of Task Similarity and the Role of Skill." *Psychological Research* 54, no. 1 (1992): 38–43.

Gobet, F. "A Pattern-Recognition Theory of Search in Expert Problem Solving." *Thinking and Reasoning* 3, no. 4 (1997): 291–313.

Gobet, F. "Some Shortcomings of Long-Term Working Memory." *British Journal of Psychology* (Pt. 4) (2000): 551–570.

Gobet, F., and G. Clarkson. "Chunks in Expert Memory: Evidence for the Magical Number Four . . . or Is It Two?" *Memory* 12, no. 6 (2004): 732–747.

Gobet, F., and H. A. Simon. "Expert Chess Memory: Revisiting the Chunking Hypothesis." *Memory* 6, no. 3 (1998): 225–255.

Gobet, F., and H. A. Simon. "Recall of Random and Distorted Chess Positions: Implications for the Theory of Expertise." *Memory and Cognition* 24, no. 4 (1996): 493–503.

Gobet, F., and H. A. Simon. "Recall of Rapidly Presented Random Chess Positions Is a Function of Chess Skill." *Psychonomic Bulletin and Review* 3(2) (1996): 159–163.

Gobet, F., and H. A. Simon. "Templates in Chess Memory: A Mechanism for Recalling Several Boards." *Cognitive Psychology* 31, no. 1 (1996): 1–40.

Gobet, F., and A. J. Waters. "The Role of Constraints in Expert Memory." *Journal of Experimental Psychology, Learning, Memory, and Cognition* 29, no. 6 (2003): 1082–1094.

Goldin, S. E. "Effects of Orienting Tasks on Recognition of Chess Positions." *American Journal of Psychology* 91 (1978): 659–671.

Golombek, H. *A History of Chess*. London: Routledge & Kegan Paul, 1976. A popular book crafted with much carefulness and caring. A real gem.

Holding, D. H. *The Psychology of Chess Skill*. Hillsdale, NJ: Erlbaum, 1985.

Holding, D. H. "Theories of Chess Skill." *Psychological Research* 54 (1992): 10–16.

Holding, D. H., and R. I. Reynolds. "Recall or Evaluation of Chess Positions as Determinants of Chess Skill." *Memory and Cognition* 10, no. 3: 237–242.

Hooper, D., and K. Whyld. *The Oxford Companion to Chess*. Oxford: Oxford University Press, 1992. A comprehensive encyclopedia with general and some detailed information about chess as an intellectual activity. Includes biographical notes about the most important players up to the present day.

Horgan, D. D. "Children and Chess Expertise: The Role of Calibration." *Psychological Research* 54, no. 1 (1992): 44–50.

Kotov, R. *Piense como un gran maestro*. Madrid: Editorial Ricardo Aguilera, 1982. A popular and idiosyncratic classic about how to calculate variations. Contains a series of interesting practical suggestions and explores some aspects of the psychology of errors in chess.

Krogius, N. V. *La psicología en ajedrez*. Barcelona: Ediciones Martínez Roca, 1979. A good book about the practical aspects of the psychology of chess.

Larkin, J. H., J. McDermot, D. P. Simon, and H. A. Simon. "Expert and Novice Performance in Solving Physics Problems." *Science* 208 (1980): 1335–1342.

Lasker, E. *Lasker's Manual of Chess*. New York: Dover, 1960. A good manual for learning how to play chess. Contains references to Lasker's philosophy based on the concept of struggle as a driving force of existence.

Lassiter, G. D. "The Relative Contributions of Recognition and Search-Evaluation Processes to High-Level Chess Performance: Comment on Gobet and Simon." *Psychological Science* 11, no. 2 (2000): 172–173, discussion 174.

Liljequist, R., and M. J. Mattila. "Effect of Physostigmine and Scopolamine on the Memory Functions of Chess Players." *Medical Biology* 57, no. 6 (1979): 402–405.

Mireles, D. E., and N. Charness. "Computational Explorations of the Influence of Structured Knowledge on Age-Related Cognitive Decline." *Psychology of Aging* 17, no. 2 (2002): 245–259.

Newell, A., and H. A. Simon. "An Example of Human Chess Play in the Light of Chess Playing Programs." *Progress in Biocybernetics* 2 (1965): 19–75.

Nimzowitsch, A. *Mi sistema: Método de enseñanza*. Madrid: Editorial Fundamentos, 1987. One of the fundamental books on strategy, written with humor but poorly translated into Spanish.

Pfau, H. D., and M. D. Murphy. "Role of Verbal Knowledge in Chess." *American Journal of Psychology* 101 (1988): 73–86.

Polgar, S., and P. Truong. *Breaking Through: How the Polgar Sisters Changed the Game of Chess*. London: Everyman Chess, 2005.

Reingold, E. M., N. Charness, M. Pomplun, and D. M. Stampe. "Visual Span in Expert Chess Players: Evidence from Eye Movements." *Psychological Science* 12, no. 1 (2001): 48–55.

Reingold, E. M., N. Charness, R. S. Schultetus, and D. M. Stampe. "Perceptual Automaticity in Expert Chess Players: Parallel Encoding of Chess Relations." *Psychoneurology Bulletin Review* 8, no. 3 (2001): 504–510.

Reti, R. *Masters of the Chessboard*. New York: Dover, 1976. A gem in the history of chess. Reti analyzes the game and different styles and provides lucid brushstrokes about the meaning of the game.

Reti, R. *Modern Ideas in Chess*. London: Hardinge Simpole, 2002.

Reynolds, R. I. "Search Heuristics of Chess Players of Different Calibers." *American Journal of Psychology* 95, no. 3 (1982): 383–392.

Richman, H. B., F. Gobet, J. J. Straszewski, and H. A. Simon. "Perceptual and Memory Processes in the Acquisition of Expert Performance: The EPAM Model." In K. A. Ericsson, ed., *The Road to Excellence* (167–187). Mahwah, NJ: Erlbaum, 1996.

Robbins, T. W., E. J. Anderson, D. R. Barker, A. C. Bradley, C. Fearnyhough, R. Henson, and S. R. Hudson. "Working Memory in Chess." *Memory and Cognition* 24, no. 1 (1996): 83–93.

Rowson, J. *The Seven Deadly Chess Sins*. London: Gambit, 2000.

Saariluoma, P., amd V. Kalakoski. "Apperception and Imagery in Blindfold Chess." *Memory* 6, no. 1 (1998): 67–90.

Saariluoma, P, and T. Laine. "Novice Construction of Chess Memory." *Scandinavian Journal of Psychology* 42, no. 2 (2001): 137–146.

Saariluoma, P. "Aspects of Skilled Imagery in Blindfold Chess." *Acta Psycholoigica (Amsterdam)* 77, no. 1 (1991): 65–89.

Saariluoma, P. "Chess Players' Intake of Task-Relevant Cues." *Memory and Cognition* 13, no. 5 (1985): 385–391.

Saariluoma, P. "Chess Players' Recall of Auditorily Presented Chess Positions." *European Journal of Cognitive Psychology* 1 (1989): 309–320.

Saariluoma, P. "Error in Chess: The Apperception-Restructuring View." *Psychological Research* 54, no. 1 (1992): 17–26.

Saariluoma, P. "Location Coding in Chess." *Quarterly Journal of Experimental Psychology* 47A (1994): 607–630.

Saariluoma, P. "Visuospatial and Articulatory Interference in Chess Players' Information Intake." *Applied Cognitive Psychology* 6 (1992): 77–89.

Saidy, A. *La batalla de las ideas en ajedrez*. Barcelona: Ediciones Martinez Roca, 1973. An analysis of the chess player personality of the great players from Botvinnik to Fischer.

Schneider, W., H. Gruber, A. Gold, and K. Opwis. "Chess Expertise and Memory for Chess Positions in Children and Adults." *Journal of Experimental Child Psychology* 56, no. 3 (1993): 328–349.

Schultetus, R. S., and N. Charness. "Recall or Evaluation of Chess Positions Revisited: The Relationship between Memory and Evaluation in Chess Skill." *American Journal of Psychology* 112, no. 4 (1999): 555–569.

Scurrah, M. J., and D. A. Wagner. "Cognitive Model of Problem-Solving in Chess." *Science* 169, no. 3941 (1970): 209–211.

Shahade, J. *Chess Bitch: Women in the Ultimate Intellectual Sport*. Los Angeles: Siles Press, 2005.

Simon, H. *Models of Thought*. New Haven: Yale University Press, 1979. A monumental book with a selection of the most relevant articles from Herbert Simon, who won a Nobel Prize in economics. A fundamental work that gathers different models and approximations to problem solving, expert dynamics, and creativity, with special attention to chess as a working laboratory. Includes classic articles written with Chase ("Perception in Chess" and "The Mind's Eye in Chess").

Simon, H. A., and W. G. Chase. "Skill in Chess." *American Scientist* 61 (1973): 393–403.

Simon, H. A., and K. Gilmartin. "A Simulation of Memory for Chess Positions." *Cognitive Psychology* 5 (1973): 29–46.

Soltis, A. *The Inner Game of Chess*. New York: McKay, 1994. An interesting trip through different aspects of the mentality of a chess player.

Spielmann, R. *The Art of Sacrifice in Chess*. New York: McKay, 1972. The basic elements of combinations and sacrifices. A joy.

van der Maas, H. L., and E. J. Wagenmakers. "A Psychometric Analysis of Chess Expertise." *American Journal of Psychology* 118, no. 1 (2005): 29–60.

Vicente, K. J. "Memory Recall in a Process Control System: A Measure of Expertise and Display Effectiveness." *Memory and Cognition* 20, no. 4 (1992): 356–373.

Vicente, K. J., and J. H. Wang. "An Ecological Theory of Expertise Effects in Memory Recall." *Psychological Review* 105 (1998): 33–57.

Waters, A. J., F. Gobet, and G. Leyden. "Visuospatial Abilities of Chess Players." *British Journal of Psychology* 93 (Pt. 4) (2002): 557–565.

Watkins, M.J., D. R. Schwartz, and D. M. Lane. "Does Part-Set Cuing Test for Memory Organization? Evidence from Reconstructions of Chess Positions." *Canadian Journal of Psychology* 38, no. 3 (1984): 498–503.

Chapter 5

Anantharaman, T., M. Campbell, and F. Hsu. "Singular Extensions: Adding Selectivity to Brute-Force Searching." *Artificial Intelligence* 43 (1990): 99–109.

Baxter, J., A. Tridgell, and L. Weaver. "Knightcap: A Chess Program That Learns by Combining TD(Đ) with Minimax Search." Technical report, Department of Systems Engineering, Australian National University, 1997.

Botwinnink, M. M. *Computers in Chess: Solving Inexact Search Problems*. New York: Springer Verlag, 1984. Hard to digest.

Campbell, M., J. Hoane Jr., and F.-H. Hsu. "Deep Blue." *Artificial Intelligence* 134 (2002): 57–83.

Campbell, M. S., and T. A. Marsland. "A Comparison of Minimax Tree Search Algorithms." *Artificial Intelligence* 20 (1983): 347–367.

Donninger, C., and U. Lorenz. "The Chess Monster Hydra." In J. Becker, M. Platzner, and S. Vernalde, eds., *FPL 2004, LNCS 3203* (927–932). Berlin: Springer-Verlag.

Feigenbaum, E. A., and J. Feldman, eds. *Computers and Thought*. New York: McGraw-Hill, 1963. Compilation of classic articles from the beginnings of artificial intelligence. Includes articles that are required reading from Turing, Newell, Chase, and Simon.

Frey, P. E., ed. *Chess Skill in Man and Machine*. New York: Springer-Verlag, 1977. A fundamental book on the development of ideas in chess programming from the

1950s through the late 1970s. Contains articles about cognition and the difference between machine and human thinking, including the following: R. M. Church and K. W. Church, "Plans, Goals, and Search Strategies for the Selection of a Move in Chess" (131–56); D. J. Slate and L. R. Atkin, "Chess 4.5: The Northwestern University Chess Program" (82–118); D. E. Wilkins, "Using Chess Knowledge to Reduce Search" (211–242).

Gherrity, M. "A Game-Learning Machine." Doctoral thesis, University of California at San Diego, California, 1993.

Gobet, F., and P. Jansen. "Towards a Chess Program Based on a Model of Human Memory." In H. J. van den Herik, I. S. Herschberg, and J. W. Uiterwijk, eds., *Advances in Computer Chess 7* (35–60). Maastricht: University of Limburg Press, 1994.

Hsu, F.-H. *Behind Deep Blue: Building the Computer That Defeated the World Chess Champion*. Princeton: Princeton University Press, 2002. A (rather triumphal) narration of the experiences of the matches between Kasparov and Deep Blue from one of the fathers of the machine. Compare with Khodarkovsky and Shamkovich, below.

Hsu, F.-H. "IBM's Deep Blue Chess Grandmaster Chips." *IEEE Micro* 19, no. 2 (1999): 70–81.

Kendall, G, and G. Whitwell. "An Evolutionary Approach to the Tuning of a Chess Evaluation Function Using Population Dynamics." In *Proceedings of the 2001 IEEE Congress on Evolutionary Computation Seoul, Korea* (995–1002). IEEE Press, 2001.

Khodarkovsky, M., and L. Shamkovich. *A New Era: How Garry Kasparov Changed the World of Chess*. New York: Ballantine Books, 1997. A narration by a member of Kasparov's team of his encounters with Deep Blue in 1996 and 1997. Also analyzes encounters between Kasparov and Anand before and after Deep Blue. Compare with Hsu (2002), above.

Levinson, R. A., and J. Amenta. "MORPH: An Experience-Based Adaptive Chess System. A Demonstration Report." *International Computer Chess Association Journal* 16, no. 1 (1993): 51–53.

Levinson, R. A., and R. Weber. "Chess Neighborhoods, Function Combination and Machine Learning." In T. A. Marsland and I. Frank, eds., *Computers and Games: Lecture Notes in Computer Science* (133–150). Berlin: Springer-Verlag, 2001.

Levy, D., and M. Newborn. *How Computers Play Chess*. New York: Computer Science Press and Freeman, 1991. A fundamental introduction to chess programming. Written by two of the most direct protagonists. Essential.

Marsland, T. A. "The Anatomy of Chess Programs." In *Papers from the 1997 AAAI Workshop. Deep Blue versus Kasparov: The Significance for Artificial Intelligence* (24–26). AAAI Press, 1997.

Marsland, T. A. "Relative Efficiency of Alpha-Beta Implementations." In A. Bundy ed., *Proceedings of the International Joint Conference on Artificial Intelligence, Karlsruhe* (763–766). William J. Kauffman, 1983.

Marsland, T. A. "A Review of Game-Tree Pruning." *Journal of the International Computer Chess Association* 9, no. 1 (1986): 3–19.

Marsland, T. A., and Y. Björnsson. "From Minimax to Manhattan." In Jaap van den Herik and Hiroyuki Iida, eds., *Games in AI* (5–17). AAAI Press, 1999.

Morgenstern, O., and J. von Neumann. *Theory of Games and Economic Behavior*. Princeton: Princeton University Press, 1944. The classic book of game theory.

Pachman, L., and V. I. Kuehnmund. *Ajedrez y computadoras*. Barcelona: Ediciones Martínez Roca, 1982. Somewhat outdated basic introduction to computation and chess programming.

Pearl, J. "Scout: A Simple Game-Searching Algorithm with Proven Optimal Properties." In *Proceedings of the First Annual National Conference on Artificial Intelligence, Stanford, CA* (143–145). AAAI Press, 1980.

Pellen, L. "Octavius." http://members.datafast.net.au/lpellen/ octavius/default.htm, 2001.

Shannon, C. E. "Programming a Computer for Playing Chess." *Philosophical Magazine Series 7* 41, no. 314 (1950): 256–275.

Smet, P., G. Calbert, J. Scholz, D. Gossink, H. Kwok, and M. Webb. "The Effects of Material, Tempo and Search Depth on Win-Loss Ratios in Chess." In T. D. Gedeon and L.C.C. Fung, eds., *AI 2003, LNAI 2903* (501–510). Berlin: Springer-Verlag, 2003.

Thrun, S. "Learning to Play the Game of Chess." In G. Tesauro, D. Touretzky, and T. Leen, *Advances in Neural Information Processing Systems (NIPS) 7* (1069–1076). Cambridge: MIT Press, 1995.

van den Herik, H. J., and H.H.L.M. Donkers. "Games, Theory and Applications." In P. Van Emde Boas et al., eds., *SOFSEM 2004, LNCS 2932* (1–8). Berlin: Springer-Verlag, 2004.

Index